The Hilbert Challenge

JEREMY J. GRAY

Senior Lecturer in Mathematics
Open University

OXFORD
UNIVERSITY PRESS

OXFORD

UNIVERSITY PRESS

Great Clarendon Street, Oxford OX2 6DP

Oxford University Press is a department of the University of Oxford.
It furthers the University's objective of excellence in research, scholarship,
and education by publishing worldwide in

Oxford New York

Athens Auckland Bangkok Bogotá Buenos Aires Calcutta
Cape Town Chennai Dar es Salaam Delhi Florence Hong Kong Istanbul
Karachi Kuala Lumpur Madrid Melbourne Mexico City Mumbai
Nairobi Paris São Paulo Singapore Taipei Toykyo Toronto Warsaw

with associated companies in Berlin Ibadan

Oxford is a registered trade mark of Oxford University Press
in the UK and in certain other countries

Published in the United States
by Oxford University Press Inc., New York

A catalogue record for this book is available from the British Library

Library of Congress Cataloging in Publication Data
Data available

ISBN   0 19 850651 1

1 3 5 7 9 10 8 6 4 2

Typeset in Minion by J&L Composition Ltd, Filey, North Yorkshire
Printed in Great Britain on acid-free paper by T. J. International Ltd, Padstow

The Hilbert Challenge

'WE MUST KNOW; WE WILL KNOW'

A history of the Hilbert Problems

In 1900 David Hilbert, then 38 and one of the leading mathematicians of the time, took the opportunity presented by the International Congress of Mathematicians in Paris to pose 23 problems which illustrated his views on the future of mathematics. What problems would it be most important to solve? What was the importance of problems for the progress of mathematics? What were the best foundations for mathematics? What was the proper relationship of mathematics to science? His intentions were not merely, as he put it, to lift the veil that separates us from the future, but to help shape and direct that future. With his prestige and that of his University behind them—and Hilbert worked at the most powerful centre for mathematics in the world—the problems he posed were always likely to be at the forefront of mathematical research, and so they became.

This book is an account of those problems: what they were, why they were proposed, and how they fared. Their stories shed light on the nature of mathematics in the twentieth century, they tell us about its conflicting priorities, they tell us what its leading figures wanted to bring about, and also what they missed. Some were a source of inspiration, others held out until long after Hilbert died in 1943. Some were entirely precise, others almost programmatic. Some were so difficult and valuable that reputations could be made by solving them, or even cracking open a part of one, while a few evaporated with their solution.

Hilbert remained an active mathematician for many years, and the history of the problems is tied up with his subsequent career. In the 1920s his interests shifted heavily towards the foundations of mathematics, and the problems he had raised that pointed in that direction acquired a new significance. In 1900 he had merely asserted with his characteristic optimism that in mathematics every problem had its solution, there was nothing we could not know. Now he wanted to show that mathematics could be proved to be correct. In 1900 his own optimism meshed with the confidence of the period, but in the 1920s it clashed with a new cultural pessimism. Moreover, he had rivals, and he sought to defeat them by proving something astonishing: that every mathematical problem is solvable in principle, and indeed can be decided in a finite number of logical steps. This was not to be, and the Hilbert Programme went down to defeat, decisively criticised by Gödel, Turing, and others.

Hilbert also gave considerable thought to the proper relation of mathematics to physics. His close friend Hermann Minkowski was a spur to his interest, and after Minkowski's unexpected death in 1909, Hilbert worked even more actively in that area, lecturing on it frequently, and getting drawn in to the search for the most general equations for physics when Einstein was struggling to create his general theory of relativity. Several of the Problems bear on the theme of giving physicists the right sort of mathematics. Hilbert not only proposed a set of potential results, he offered axiomatic mathematics as an organising principle that would impose a clear logical structure on every branch of physics. Most physicists found that their subject, which soon plunged into the disorienting swirl of quantum mechanics, moved too fast for them to be able to heed Hilbert's advice, but mathematicians continued to work through the mathematics appropriate to physics, often in Hilbertian ways and often with great success. In a century where some have castigated mathematics for being too abstract and losing its way by losing touch with the sciences, and have blamed Hilbert among others, Hilbert's life-long stress on the importance of that link actually offers a very different message, and a different defence for mathematics.

Hilbert was, of course, a pure mathematician, many would say one of the greatest. One part of the subject he cherished most was the theory of numbers, and it afforded him roughly a quarter of the Problems he proposed. But even here he was far from complacent, and keen to persuade others of the deep significance of the topic. Indeed, one key to the success of the Problems down the years is that he anchored them in a context, and gave other mathematicians good reasons to care about them. Problems, Hilbert believed, were the life of the subject. If a topic generates no problems then it is dead. By following the history of the Problems we can trace the life of mathematics, with its uncertainties, false starts, successes, and transformations.

The story of the Hilbert Problems is also a story of status. How are priorities set in mathematics? Who sets them, who disagrees? How are these disagreements handled? How is a reputation acquired, and what does it bring in its wake? Status in mathematics accrues not only to people and to universities, but to topics and problems. It can seem just and appropriate, or capricious and cliquish. The intellectual and the social are inter-twined, and one hundred years after Paris is as good a time as any to lift the veil that now separates us from the past, and see how the future of mathematics was created.

Mathematics can be read on several levels, and of course some are accessible only to the expert. But we can always ask, and we should always ask: what is important about that? What is interesting about that? And we shall see that even the expert is not content with a long, technical argument. A conceptual account that explains why some theorem is true is preferred to a calculation that can only be followed line by weary line. Hilbert was quite clear about this, and in this way he was echoing, and amplifying, the view of many mathematicians of his time. His friend Minkowski had urged mathematicians 'to face problems with a minimum of blind calculation, a maximum of seeing thought.'[1]

I have tried in this book to pursue the mathematics with an eye to its importance and its interest. I hope it will be clear when, in the opinion of this or that mathematician, a topic was significant, and why. I have thought it better to leave some of the details in, rather

---

[1]   Minkowski on Dirichlet, quoted in Weyl (1946).

than reduce the story to one in which this or that mathematician joins what has been called 'the honours class of the mathematical community' by solving a problem I have not adequately described with methods I could not bring myself to mention.[2] If the discussion takes you, the reader, into some thorny points it is always possible to leap over to the next section. As one can guess, the level of difficulty rises slightly as the book proceeds—it is the harder problems, after all, that held out longest—and I have tried to put the more difficult topics towards the end of the chapters.

Problems on the foundations of mathematics and analyses of why, or even if, mathematics is true naturally have a claim on our attention. So too do contributions to making mathematics useful. Nonetheless about a half of Hilbert's Problems concern questions in pure mathematics. This reflects more than a century's agreement that this was right—the heart of mathematics is pure. We shall see in Chapter 6 that this view has been contested, and throughout this book that it has seldom been held without qualification. Hilbert held it himself. It is the purest problems, roughly numbers 7 to 18, that may seem the most difficult to appreciate. I would like to suggest an approach to them here which emphasises that the basic objects mathematicians would like to know about are numbers (indeed the positive integers 1, 2, 3, . . .) and shapes, including curves. They can be introduced and studied in all sorts of ways: by algebra, by analysis, by axioms. It is particularly interesting to study how they can be transformed, and this is done mathematically by introducing all sorts of functions, which can also be understood in all sorts of ways. But the fundamental nature of numbers and shapes is clear and apparent. All of Problems 7 to 18 concern them, some in ways that addressed traditional questions and for which there was already a body of literature, some in quite novel ways. Taken together, they provide a remarkable range of examples of what mathematicians did not know in 1900. One approach to the accounts that follow is to savour how easy it is to encounter not a sea of confusing formulae in mathematics but a wealth of good questions about our ignorance. What mathe-

---

[2] Weyl (1944), in (1968) 4, p. 136.

maticians did with Hilbert's Problems was profound, but it was not perverse. To appreciate that is to rescue mathematics from the arcane and to see it as one of the most remarkable achievements of the human mind.

Acknowledgements

It is a pleasure to thank numerous people who have answered my questions, corrected my mistakes, and generally helped this book to come about, in particular:

Tom Archibald, June Barrow-Green, Felix Browder, Roger Cooke, Leo Corry, Moritz Epple, John Fauvel, Matt Frank, Catherine Goldstein, Moshe Machover, Bill Parry, David Rowe, Erhard Scholz, George Smith, K.-G. Steffens, Ian Stewart, Jamie Tappenden, Mrs Gonska for producing the electronic copy of Hilbert's lecture, and Kathleen Whalen the Manuscripts Librarian at Bryn Mawr. I was helped by the invitation to address the Mathematics Association of America Mathfest in July 1999, and would like to thank them for their comments on the work in progress.

Jeremy Gray
Open University

Contents

CHAPTER 1

The future unveiled

On Wednesday 8 August 1900, David Hilbert stepped up to the rostrum at the International Congress of Mathematicians in Paris to give his lecture on Mathematical Problems. All lectures at conferences are pieces of theatre. The audience has some expectations, the lecturer some hopes. But on this occasion the hopes and expectations ran particularly high, because Hilbert was emerging as the leading mathematician of his generation, and his theme promised to be broader and more exciting than the usual technical accounts of mathematics, however successful.

'Who among us', he began, 'would not be glad to lift the veil behind which the future lies hidden; to cast a glance at the next advances of our science and at the secrets of its development during future centuries? What particular goals will there be toward which the leading mathematical spirits of coming generations will strive? What new methods and new facts will the new centuries disclose in the wide and rich field of mathematical thought?'

With these ambitious words Hilbert seized his moment. As we shall see, he had prepared for this for a year, trying out drafts of his talk on his friends, selecting the topics that seemed to him to be most fruitful. He had the authority of an expert to attempt such a task, and he knew very well that by lending his authority to whatever problems he chose he would attract people to work on them. But he did more. He was not the only person, as one century ended and another began, to play at making predictions, to promote their favourite subjects. But he had perhaps thought harder than most people about what such an exercise can best involve.

Hilbert chose problems because he believed that mathematics advanced by solving problems. Problems, as he told his audience, are a sign that the subject is alive. The benefit of problems is that, by solving them, the investigator gains a wider view of the subject. Of course some problems are better than others. The best, he said, are characterised by their clarity and ease of comprehension, virtues that are more commonly attached to theories. Problems should be difficult, but not completely inaccessible, and uncovering the successful solution should be a source of pleasure.

Thus far Hilbert had only offered the satisfaction of good honest work. He next connected the worth of problems to the theories that they inspire. This was to prove a key step in ensuring the longevity of the problems, because it gave the aspirant problem-solver an extra reason to care about whichever problem they picked. A first-rate problem, he implied, is one that opens up a subject, and perhaps leads to a new field of research. It will in that way live on even if it is in some sense solved.

He gave the example of the famous problem, originally set by Johann Bernoulli in 1696, which asks for the curve in a vertical plane that joins two points and which is the curve along which a particle would slide from the one point to the other in the shortest time. It is sometimes called for that reason the curve of quickest descent. The answer was promptly discovered by several mathematicians to be a part of a catenary (see Box 1.1); Bernoulli himself recognised Newton's solution at once although it was submitted anonymously 'As the lion is recognised by his print', he said.[1] Hilbert's audience knew the history of the problem. They also knew that it had eventually generated a profound and varied set of similar problems, and that the accompanying theory was much used in physics. To give another example, light travels along the curve of least time between any two points. The major physics textbooks of the late nineteenth century were full of similar principles, known by such names as the method of least action, which asserted that such-and-such a physical process would always happen in such a way that some quantity was minimised (or, less often, maximised). So the problem of finding the curve of quickest descent was solved, but the problem area was still full of life. Indeed, the

---

[1]   Westfall (1983) p. 583.

Box 1.1 The brachistochrone problem

It is required to slide a bead down a wire from a point A to a point B in the shortest possible time. The bead has a given initial velocity. What should be the shape of the wire?

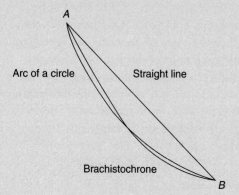

The answer is the cycloid through A and B which has a vertical tangent at A. The initial data determines the curve uniquely. Descent along this curve is faster than along the straight line through A and B and through the appropriate arc of a circle (Galileo's candidate).

experts in the audience would have known that there were substantial mathematical problems here. The confidence of the physicists was not being matched by the ability of the mathematicians to come up with theorems.

Hilbert then switched to Fermat's Last Theorem.[2] This is the celebrated claim that the equation $x^n + y^n = z^n$ has no solution in non-zero integers whenever n is an integer greater than 2. In 1900 this was known for quite a range of exponents n, including all but three prime values of n less than 100, but the problem was far from being solved. It had however generated a large industry exploring

---

[2] Finally solved by Andrew Wiles, *Annals of Mathematics*, 1995.

the vast new domain of algebraic number theory, the discipline Hilbert had already refashioned in 1897. His *Report on the Theory of Numbers*[3] was the work that had consolidated Hilbert's reputation, so it is no surprise to see him picking a problem from this area. Again, however, the implication was that the original problem was interesting because of the subject it had opened up. Fermat's Last Theorem was not solved, but Hilbert was not as interested in it as in the whole slew of questions and ideas that formed the subject of algebraic number theory.

Hilbert was not content to observe that problems can lead to theories that in turn presumably generate problems, whether in applied mathematics or pure mathematics. He wanted to argue that this process breaks down, or transcends, the division that had been building up between pure and applied mathematics. Problems can find the most unexpected significance. The question of the curve of quickest descent comes up in mechanics, in a branch of analysis called the calculus of variations, and more recently in the foundations of geometry. Felix Klein, said Hilbert—in what was not just a politic allusion to his Head of Department—had shown how the regular or Platonic solids mattered in such diverse fields as elementary geometry, group theory, the solution of polynomial equations, and the solution of some ordinary differential equations.[4]

Mathematics picks up some of its problems from the real world. But, he said, it has been the modern experience that self-reflection and criticism also generates valuable problems: on the one hand, problems such as Bernoulli's, and on the other, Fermat's. The surprising and fruitful analogies between different questions, methods, and ideas that the mathematician frequently observes originate, Hilbert suggested, in the ever-recurring interplay between thought and experience. In this way mathematics was not, for Hilbert, a collection of different things but a coherent, dynamic, evolving whole.

---

[3] Usually known by the opening words of its German title as the *Zahlbericht*. Hilbert discussed Fermat's Last Theorem briefly in its concluding pages.

[4] Klein (1884).

Box 1.2 The regular solids; their symmetries form a group

The five regular solids are the tetrahedron, the cube, the octahedron, the dodecahedron and the icosahedron. They are the only figures all of whose faces look the same, all of whose vertices look the same, and which can be inscribed in a sphere.

The symmetries of a regular solid provide a good example of a group. Consider the cube. It can be picked up and put back where it was in a number of ways. Any of these form a symmetry of the figure. The result of following one symmetry with another is to produce a third symmetry, which is the key property of a group.

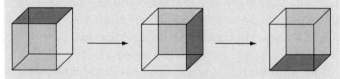

How many symmetries of the cube are there? Pick a face, and two adjacent vertices on it (marked A and B in the figure). This face can be put on top of any one of 6 faces.

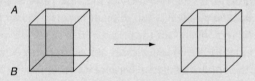

The vertex A can go in any one of 4 positions. This gives a total of 6.4 = 24 ways of putting the cube back in place, and that is all there are unless you allow reflections as well, in which the cube is replaced by its mirror image. If reflections are allowed there are 48 symmetries.

Hilbert continued to push his argument. For a problem to be solved, he said, there should be an argument starting from a finite number of exactly formulated hypotheses and leading, by a finite number of deductions, to the conclusion. This ensures the rigour of the solution method. Rigour, however, need not be disparaged (as it sometimes is, even by mathematicians). Rigour can be the way in which clarity is obtained, and the theory simplified, he said, and he held out the example, or the hope, of the calculus of variations. Nor was rigour the exclusive property of certain parts of mathematics, such as arithmetic and analysis. In fact, geometry, mechanics, even physics could be made completely rigorous. This is an astonishing claim. Geometry had been declining throughout the nineteenth century as a paradigm of rigour, even though it was an abundant source of problems and theories. The mathematicians' sense of rigour had shifted onto other topics, notably arithmetic, and Hilbert had particularly strong views on what that entailed. Mechanics, and still more physics, had never been thought of as rigorous according to the high standards Hilbert had just laid down. His remarks here were deliberately provocative, setting the scene for some of his later proposals.

Finally, Hilbert turned to offer some advice about how difficult problems can be solved. It might be that the view point of the mathematician is insufficiently general, and from within a more general theory the problem can be seen as a tractable special case. Or, and perhaps this is more often the case, specialisation is required, and the problem is too difficult because simpler special cases are still too difficult for the researcher. In such cases nothing can be done until the simpler cases are solved. Sometimes, he suggested, the problem is elusive because it is not properly formulated, and when it is properly put it becomes possible to show that it actually cannot be solved. Thus the ancient Greeks discovered that there is no rational number which is the square root of 2. But a properly formulated problem and accompanying theory still lead to a solution, in the form of a rigorous proof of the impossibility of the original task. In the same way (switching to applied mathematics, as he had indicated was natural) the failure to construct a perpetual motion machine led to an analysis of why such machines are impossible and to the discovery of the law of conservation of energy. Such successes often inspire mathematicians to believe that every problem is capable of solution; the enterprise is not hopeless in principle.

Then came the famous peroration, often quoted later. 'This conviction of the solvability of every mathematical problem is a powerful incentive to the worker. We hear within us the perpetual call: There is a problem. Seek its solution. You can find it by pure reason, for in mathematics there is no *ignorabimus* (we shall not know).'

After this fine flight of rhetoric, so thoroughly in tune with the self-confidence of the age, and so deliberately opposed to the end-of-century pessimists, Hilbert had to deliver problems that exemplified his principles. In his speech he offered 10, but he also distributed the full text of his speech, rather astutely in French because Hilbert spoke in German, and there he presented the full list of 23 problems.

The 10 Problems that carried the weight of his talk as a theatrical occasion began with Cantor's problem of the cardinal number of the continuum. This asks: How many real numbers are there? The answer is to be given in terms of the German mathematician Georg Cantor's recent theory of infinite sets. The suggestion was that the real number continuum is the smallest infinite set that is strictly larger than a countable set. Cantor had already shown that the set of all real numbers was not countable, but proving that they form the smallest non-countable set had so far defeated everyone.

Next came the problem of proving that the axioms for arithmetic are consistent. This might seem peculiar. It is hardly likely that counting and primary school arithmetic are suddenly going to turn out to be fundamentally flawed. But how would you establish this to Hilbert's satisfaction? To do that, the mathematician must furnish exact hypotheses and then deduce the consistency of the axioms in a finite number of steps. Without a rigorous argument, some element of clarity is missing. Mathematicians, who had had to learn that Euclidean geometry is not the only possible geometry, could be forgiven if they wished to subject arithmetic, that other intuitive pillar of mathematical knowledge, to rigorous examination. Moreover, Hilbert's own recent analysis of several types of geometry had shown that each taken separately is consistent by exhibiting examples that satisfy the appropriate axioms, and those examples were couched in terms of arithmetic. So the geometry

Box 1.3 Countable and uncountable

A set is said to be countable if it can be put into a 1–1 correspondence with the natural numbers, thus labelling every
element with a unique positive integer and assigning an
element to each positive integer.

For example, the square numbers are countable:

$$\begin{array}{ccccccc} 1 & 2 & 3 & 4 & & n \\ \updownarrow & \updownarrow & \updownarrow & \updownarrow & \cdots & \updownarrow \\ 1 & 4 & 9 & 16 & & n^2 \end{array}.$$

So too are the positive rational numbers. This is best done by
arranging them in an array, striking out all duplicates (such as
$1/3 = 2/6 = 3/9 = \cdots$) and then running through the rest as
shown, counting as you go:

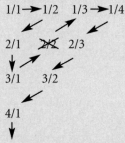

It was one of Cantor's remarkable discoveries that there are
sets which are not countable. The real numbers between 0 and
1 form such a set, for, as he showed, if they were countable one
could derive a contradiction as follows. Let us enumerate them
in a list, with some convention about how to avoid repetitions
caused by repeating nines (for example $0.012 = 0.01199999$
$\ldots.$). Let us look at the first digit of the first number, the
second digit of the second number, and so on, so we look at the
nth digit of the nth number, and write down a new number in
this way. If the digit we are looking at is 5, write down 2. If it is
not a 5, write down 5. Now consider the real number we have
just written down. It is not in our list, because if it were it
would be the nth number for some n (as it might be, the

2075th). But it differs from the nth digit in the nth decimal place. So it is not in our list. But we supposed our list contained all the real numbers between 0 and 1. From this contradiction it follows that the real numbers are not countable.

makes sense if the arithmetic does, and Hilbert was now asking that the arithmetic be shown to be intellectually coherent.

Next Hilbert called for an axiomatisation of physics. This would give successive branches of physics foundations that did not contradict each other, and Hilbert shrewdly suspected that this was not always the case. Instead, as he observed on other occasions, physicists were in the habit of enshrining as axioms whatever seemed to them to be a sufficiently profound insight, or assumption, without regard for consistency. This made it very hard to know what claims would fall whenever a trenchant new experiment came along.

Then came some problems from number theory. The problems he used to illustrate his theme were chosen with a good eye to his audience. In 1873 Charles Hermite, the leader of the French mathematicians, had shown that the number e, the base of the natural logarithms, is transcendental. This means that it is not the root of any polynomial equation of any degree with integer coefficients. Ten years later Hilbert's own supervisor, Lindemann, had shown that π was also transcendental, and since then a number of mathematicians, Hilbert among them, had simplified Lindemann's original proof until Klein could claim that 'the transcendency of e and π should henceforth be introduced into university teaching everywhere.'[5] The numbers π and e are among the most useful and fundamental in all mathematics, so these theorems were profound illustrations of the conceptual precision of mathematics at the dawn of the twentieth century.

---

[5] Klein (1893) p. 53. I am indebted to Roger Cooke for pointing out to me that Lindemann's original proof was one Weierstrass found less than satisfactory, so the later versions are not just simpler but also more rigorous. This provides a small illustration of the validity of Hilbert's insistence on rigour.

True to his faith that good problems have deep roots in the subject, Hilbert observed that something very odd is going on with π and e. The famous equation $e^{i\pi} = -1$ makes the point. If e and π are transcendental numbers, how extraordinary it is that such a simple combination of them should be an integer. On the face of it, the thing to expect is that $e^{i\pi}$ is a truly awful number about which nothing interesting can be said. In order to explore further, and to do so with a definite purpose, Hilbert therefore proposed the problem that 'The expression a^β, for an algebraic base a and an irrational algebraic exponent β, e.g., the number $2^{\sqrt{i}}$ or e^π [which is equal to] i^{-2i}, always represents a transcendental or at least an irrational number.' So if raising a number to a power gives a rational number, then either the matter is trivial (as it is with such equations as $\sqrt{2}^2 = 2$) or one of the numbers is a transcendental number.

Hilbert then alluded to the fact that Riemann, the great German mathematician of the mid-century, had published a paper in 1859 in which some deep and technical properties of a function (called the Riemann zeta function) were shown to have significant implications for the distribution of the prime numbers. These suggestions by Riemann surely put the specific problem of the distribution of the prime numbers into the right general context, but it was still holding out. There had been some real progress, as Hilbert mentioned, by a young French mathematician, Jacques Hadamard, and a Belgian, de la Vallée Poussin, who had independently established one of the main results that Riemann had rather skipped over, and that would have contributed to the feeling that the Riemann zeta function was the key.[6]

He also alluded to Goldbach's conjecture that every even number greater than 2 is the sum of two prime numbers, which he thought might come after a successful attack on the prime number theorem. At all events it has not come before: the best result so far is

---

[6] Attempts on the Riemann hypothesis would fill a difficult book, and it is still unresolved. The reader is therefore referred to Patterson (1988). For a history, see Edwards (1974). Hadamard's contribution is well described in Maz'ya and Shaposhnikova (1998).

Vinogradov's result that every sufficiently large odd number is the sum of three primes.[7]

Next Hilbert took up a question that he drew from the work of the French mathematician Maurice d'Ocagne, and after that he asked a question about the shape of curves and surfaces defined by polynomial equations. He seems to have liked these questions. He returned to the first one some years later, and the second one had been the theme of a paper in 1891, but later generations were not to be so captivated. He ended, however, with a trio of problems that many would have said were central to the mathematics of the day, and indeed the century.

He picked up the thread that led back to Bernoulli's problem, and asked what might seem a technical question: does a certain class of problems in the calculus of variations (the 'regular' problems) always have analytic solutions? Removed from its technical dress, this asked if the kind of problems that arise in physics as problems in the calculus of variations must have solutions that can be expressed in terms of the most well-behaved kind of function. Mathematicians had examples of problems in the calculus of variations that did not have such nice solutions, so Hilbert was asking for a characterisation of the problems that have well-behaved solutions which is broad enough to encompass the physical ones and precise enough to permit a proof that the solutions will be analytic. Show, in other words, that every problem of this kind in physics has a solution of the kind the physicists want and expect.

Next came a much more technical problem, but one with an impeccable pedigree. It brought together the work of the German mathematician Lazarus Fuchs on linear ordinary differential equations with contributions of both Klein and the great French mathematician Henri Poincaré. As Hilbert had observed, Klein had fruitfully exploited the fact that from such a differential equation one can obtain a group. The problem was to show that given a group one can find a differential equation with the given group as

---

[7] See Baker (1984) p. 6. The phrase 'sufficiently large' means greater than some number which, unfortunately, Vinogradov's method does not permit one to determine. There are also results about even numbers, but they are harder to state.

Box 1.4 The calculus of variations

The calculus of variations can be illustrated by Fermat's principle in optics, which asserts that the path of a ray of light from a point A to a point B takes the shortest possible time. In a uniform medium this will be a straight line. In a medium where the refractive index varies, the path may be curved, and the light will then seem to come from unexpected directions (as it does in mirages).

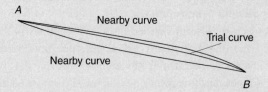

To solve a problem in the calculus of variations, a trial curve is compared with all nearby curves, those that are obtained by varying the trial curve slightly and the time along each curve is calculated (whence the name calculus of variations). The solution curve has the property that the time is longer along all nearby curves.

its group. Mathematicians like these tidy pairings: to every equation there is a group, to every group an equation. Such results suggest that the theory is well understood.

The ten problems were rounded off by one Poincaré had already tackled and almost solved. It asks if every curve described by an algebraic equation in two variables can be parameterised by suitable functions of a particular kind. It is trivial that the curve with equation $x^2 + y^2 = 1$ can be parameterised by the trigonometric functions: set $x(t) = \cos(t)$ and $y(t) = \sin(t)$. To every point of the curve there is a unique value of the variable t in the range 0 to 2π. Mathematicians had found many examples of similar parameterisations of more complicated curves, and as a result been able to solve many questions about curves. The audacious, but not rigorously established claim of Poincaré, was that any algebraic curve can be parameterised by means of new functions that he had introduced.

Hilbert ended his talk with a further peroration, urging mathematicians on to discover new and unexpected relations between hitherto separate branches of the subject. Nor need one fear that it would all become too much for the mathematician, and the subject simply become too big. Instead experience suggested that sharper tools and simpler methods would always be found to illuminate earlier theories and to cast aside older approaches. In this way mathematics would remain unified, and provide the foundation of all exact knowledge of natural phenomena.

The high drama of the lecture did not immediately excite its audience. Charlotte Scott, the British mathematician who was a Professor at Bryn Mawr, recorded that there was some desultory discussion after Hilbert finished. It was claimed that more had been done on the problem Hilbert had extracted from the work of d'Ocagne than Hilbert had suggested. The Italian mathematician Giuseppe Peano, who was the leading figure in a considerable Italian study of the foundations of geometry, suggested that more was known about the problem of axiomatizing arithmetic, and indeed that one of his associates, Alessandro Padoa, would be reporting on the topic at the Congress.[8] According to Scott, Peano said that Padoa and others had in fact solved the problem, so far as that was possible. But then the audience dispersed, and Hilbert was left to wait and see if indeed he had lifted the veil between us and the future. Or rather if by the force of his example, the quality of his arguments, and the lure of his problems, he had in fact done something to shape that future.

---

[8] Peano had proposed axioms for the natural numbers in 1889 and 1891. Hilbert's axioms for arithmetic were aimed at the real numbers, and thus focused on a different set of problems altogether.

CHAPTER 2

The shaping of a pioneer

Hilbert's student years

Much of the force of the Hilbert Problems, and their power to attract other mathematicians, derived from the reputation of their proposer, which was already considerable in 1900 and grew considerably thereafter. At 38, Hilbert was at the height of his abilities, and with a number of remarkable successes to his credit. To understand his bid to lead the mathematical profession, it is first necessary to know much more about him.

Hilbert was born in Königsberg on 23 January, 1862.[1] Königsberg was a small town in East Prussia, best known for being the home town of the philosopher Immanuel Kant, and Hilbert liked to make Kantian allusions in his work from time to time, although it can be doubted if he was ever profoundly a Kantian. Hilbert's family had lived in Königsberg for several generations, and Hilbert was proud of his origins. He was the only boy in the family, and was sent to school at the Friedrichskolleg when he was eight, but the family eventually removed him because he did not thrive there. In particular the emphasis on ancient languages and the heavy demands on the use of memory was uncongenial. He spent his last year of school at the Wilhelms-Gymnasium, and took the Abitur or school-leaving exam there. Although the new school

---

[1] I have used Blumenthal's short but highly informative biography of Hilbert that appears in Blumenthal (1935) (Blumenthal was Hilbert's first student) and Constance Reid's book (Reid, 1970), based on many reminiscences of people who knew Hilbert.

was more sympathetic to Hilbert's mathematical talent, there was no clear sign of his later eminence. Hilbert himself said that 'I did not particularly concern myself with mathematics at school because I knew that I would turn to it later.' Hilbert then went to the university in Königsberg in 1880. Although the university was small, it had a strong tradition in mathematics and physics, beginning with Carl Jacobi and the physicist Franz Neumann. Jacobi, one of the leading mathematicians of the first half of the nineteenth century, and who worked on a variety of topics from the theory of numbers to theoretical mechanics, had introduced the first mathematics seminar in a German University there, the first attempt to prepare mathematicians for a life of research. He had taken his inspiration from the situation in linguistics, which was one of the leading intellectual disciplines of the time. Franz Neumann meanwhile pioneered the teaching of experimental physics. Their efforts were successful, and the university produced a stream of eminent mathematicians and physicists, despite its size and the rise of the much larger and better located University of Berlin. Among the next generation of professors were the physicist Gustav Kirchhoff and the geometer Otto Hesse, and later Heinrich Weber held the chair in mathematics at Königsberg from 1875 to 1883.[2] Weber was succeeded by Ferdinand Lindemann, a geometer who had just become famous for his proof that π is transcendental.

Of these, the mathematician Heinrich Weber was a significant influence. He was quite broad in his interests. In 1876 he and Dedekind had published the posthumous edition of Riemann's papers, including a generous selection of unpublished material.[3] This was the major source of information for the many who came to respond to the challenges Riemann had left for future generations. In 1882 Weber published a major paper with Richard Dedekind on the algebra and geometry of algebraic curves, from a novel and abstract standpoint. He was one of the first to define an

---

[2] See Olesko (1989).
[3] Riemann (1990).

abstract group in print.[4] Through Weber, Hilbert came into contact for the first time with the strong current in German mathematical life that led back to Gauss. He took lecture courses from Weber on elliptic functions, number theory, and a seminar on invariant theory. Weber's replacement, Lindemann, also encouraged Hilbert to study invariant theory. In due course invariant theory, which we shall describe below, was to be the subject of his first great success as a mathematician, but that was to be some years away: Hilbert was not a mathematical prodigy. He also took advantage of the German University system, which allowed students to study wherever they wished, to go to Heidelberg to study for a term under Lazarus Fuchs, but otherwise Hilbert stayed at home. It is notable that he did not, for example, spend any time at the much more dynamic University of Berlin, which suggests that Königsberg was getting something right.

It was at Königsberg that Hilbert met two people who were to be lifelong influences. The first of these was Hermann Minkowski, a fellow student who was two years younger, but already a term ahead. Minkowski was a prodigy. In 1883 the Academy of Sciences in Paris announced as one of its regular prize competitions a question in the theory of numbers. Minkowski, then 19, sent in a complete answer. What was even more embarrassing for the French organisers was that the English mathematician H.J.S. Smith also wrote in to point out that he had already tackled that precise question some years before and published a complete solution. Some voices could now be heard suggesting, entirely falsely, that Minkowski had entered the competition corruptly, knowing of Smith's work. Then Smith died. The only way out for the French was to proclaim Minkowski and Smith joint winners, which they duly did. Unlike Hilbert, Minkowski had taken himself off to Berlin for a time, and through him Hilbert came to know more about the triumvirate of Kronecker, Kummer, and Weierstrass and the traditions of mathematics that they represented.

In Easter 1884 Adolf Hurwitz, who was only three years older

---

[4] Weber (1893).

Box 2.1 Elliptic functions

Elliptic functions are functions of a complex variable which generalise the trigonometric functions. The familiar sine function $y = \sin(x)$ is a function of a real variable. It takes real values and is periodic with period 2π: $y = \sin(x + 2\pi) = \sin(x)$. An elliptic function $w = \wp(z)$ is a function of a complex variable that takes complex values and has two distinct periods: $w = \wp(z) = \wp(z + \omega_1) = \wp(z + \omega_2)$. The quotient ω_1 / ω_2 is not allowed to be a real number. It follows that the elliptic function is known when its values are known on the parallelogram with vertices $0, \omega_1, \omega_2, \omega_1 + \omega_2$.

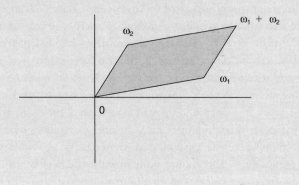

than Hilbert, was appointed to an extraordinary professorship at Königsberg. This was the first real step onto an academic career, in that there was a selection process, but it cannot be said that it was well paid. Hurwitz had studied in Berlin, where he had followed Weierstrass's lecture course on complex function theory, but he was a doctoral student of Felix Klein, then in Leipzig. In his thesis he had brought together ideas of Klein and Dedekind, and used them to re-work older ideas of yet another gifted young German mathematician, Eisenstein. Not surprisingly Hurwitz inspired Hilbert with the desire to be a universal mathematician. Hilbert recalled to Blumenthal that 'Minkowski and I were totally overwhelmed by his knowledge and we never thought we would ever come that far.'[5] It was through Hurwitz that Hilbert gained entry to the circle Felix Klein was building around him. From 1886 to 1892, when Minkowski was in Bonn, Hilbert went almost daily on mathematical walks with Hurwitz. Learning to walk and talk mathematics right across the syllabus was very important for Hilbert, and he later took the tradition with him to Göttingen. With Hurwitz, he said, he rummaged through every corner of mathematics, with Hurwitz always in the lead.

After Hilbert's graduation

Hilbert submitted his doctoral thesis in February 1885, with a thesis on invariant theory suggested by Lindemann. That May he sat the State Examination required of potential teachers in mathematics and physics. Then in the winter he set off on his student travels. First he went to Leipzig to meet Klein, and then in March he went to Paris, where he met up with his fellow German, Eduard Study. Klein had very strong positive memories of his own trip to Paris in 1870, when he went with Sophus Lie and met some of the future leaders of mathematics in France, including Camille Jordan and Gaston Darboux. His trip had been interrupted by the outbreak of the Franco-Prussian War, but nonetheless it had been a real force on his development as a mathematician, and he seems to have hoped Hilbert's visit would have a similar effect. In the event it was less successful. Study was no Lie, but a sharp-tempered critic of those who did not come up to the exacting standards he set himself, which was nearly everyone. He was similarly dismissive of nearly every branch of mathematics except the one he worked on himself, and even that had to be done his way.

Hilbert met d'Ocagne and took to him at once, and he also met Henri Poincaré, who was some eight years older than him, and had already established himself as one of the two leading men of his generation in France. But while there seems to have been mutual respect between the two figures who were in due course to lead

---

5   Blumenthal (1935) p. 390.

their mathematical nations, and on the German side some rivalry, there was no great bond of affection or coincidence of interests.

At this stage in Poincaré's career he had just left the subject of automorphic function theory. This was a remarkable unification of ideas from complex function theory, group theory, non-Euclidean geometry, and the theory of linear ordinary differential equations. Poincaré had taken it up on the occasion of a competition organised by the Academy of Sciences in Paris in 1879, which, strange to say, he did not win but came in second. But the subject had continued to grow and it had drawn him into close contact with Felix Klein, who had been exploring some of the same ideas. Something of a co-operation and something of a competition sprang up between them, which ended when Klein's health collapsed in the late summer of 1881.[6] This left Poincaré free to outline a theory in which two-dimensional non-Euclidean geometry played a central role, and which was developed with a certain amount of rigour, and a theory reliant on three-dimensional non-Euclidean geometry, which was much less precise. In 1886 Poincaré had just begun to re-direct another of his research topics, the theory of differential equations for real functions of a real variable, towards a new and prestigious prize offered by the King of Sweden, and which was on celestial mechanics and the stability of the solar system.[7] There too he was to offer a line of reasoning that mixed insight with proof, often to confusing effect.

The visionary style was very much Poincaré's, and often annoyed his colleagues. Thus Hermite once wrote to Mittag-Leffler about problems they had understanding the essay Poincaré submitted for the Swedish prize competition they were judging: 'But it must be acknowledged, in this work as in almost all his researches, Poincaré shows the way and gives the signs, but leaves much to be done to fill the gaps and complete his work. Picard has often asked him for enlightenment and explanations on very important points in his articles in the *Comptes Rendus*, without being able to obtain anything except the statement: "it is so, it is like that", so that he seems

---

[6]   See, for example, Gray (1999).
[7]   Barrow-Green (1997) and Diacu and Holmes (1996).

Box 2.2 Non-Euclidean geometry

Non-Euclidean geometry is depicted on the disc in such a way that all the white and all the black triangles are congruent, although triangles out towards the edge appear shrunk.

Compare this with the picture below of the faces of a regular dodecahedron obtained by stereographic projection from the sphere to a plane. Each of its faces has been divided into 5 black and 5 white triangles. All the triangles in the figure are congruent, but the ones further away are made to look larger by the projection.

In each case there is a perfectly valid geometry, which is not represented with complete accuracy on the plane.

like a seer to whom truths appear in a bright light, but mostly to him alone.'[8]

Visionary imprecision was very far from Hilbert's style, even though he was to make a considerable number of mistakes in his own work as the years went by. Poincaré was much more geometric, whereas at this time Hilbert's interests were heavily algebraic. This may have been a further factor keeping the two men apart, because even the best mathematicians often have temperamental preferences for one topic over another, and find their minds work more swiftly in one direction than another. Another factor may well have been some element of personal chemistry lacking in Poincaré, because, for whatever reason, he attracted almost no students and

---

[8]   Hermite to Mittag-Leffler, 22 Nov 1888, quoted in Barrow-Green (1997), p. 65.

produced very few followers whereas Hilbert, as we shall see, was not only to have a great number of students, but could count some of the best mathematicians among them. It is clear which personality would be better for building up a network of supporters.

Hilbert returned from Paris to a job in Königsberg as a *Privatdozent*, a fledgling post in the German system that was occupied by those seeking an academic position, for which the necessary and almost sufficient condition was to complete a further process, the *Habilitation*. The Habilitation gave the successful candidate the right to teach at a German University, although no-one had to pay them. Klein questioned Hilbert's decision to habilitate in Königsberg, thinking to move the younger man out into the mainstream, but to no avail. The result was that Hilbert completed the process quickly and successfully, but with a thesis topic that was not propitious for his later work. Hilbert was now able to embark on a professional career, although without a major piece of his work to his credit, and to do so at the university of his choice (his beloved Königsberg) and with Hurwitz as a colleague. Not a bad situation, but not a remarkable one either. He could count on the support of Klein, who was not only a very good mathematician but a consummate academic politician, but there were many, especially in Berlin, who distrusted Klein, and Königsberg was outside the network of Berlin professors and their well-placed former students. Nothing yet marked him out as the man to watch, unless it was Klein's prescient judgement that he was indeed the coming man.

But whatever Königsberg was making up for with Hilbert's colleagues it lacked as a teaching institution. Königsberg attracted very few students in the late 1880s and 1890s, and Hilbert, who certainly became a very good lecturer, often had an audience of only 2 or 3. On one occasion, the Winter Semester 1891/92, Hilbert had an audience of exactly one for a course on complex function theory, and that was a visiting American mathematician, Professor Franklin from Baltimore. Hilbert nonetheless seems to have taken his duties very seriously, and to have acquired the habit of lecturing on subjects that he wanted to master for the purposes of research. Later he referred to these years in Königsberg as a time of slow ripening, but at the time he knew that staying there was not sufficiently stimulating, and in Easter 1888 he set off on his travels again, going this time to Göttingen (where Klein now was) and Erlangen, to talk with Paul Gordan.

Hilbert's breakthrough in invariant theory

Hilbert was introduced to Gordan by Klein and came alive as a research mathematician. Gordan, sometimes referred to as the King of invariant theory, was a former colleague of Klein's, and although a more contrasting couple cannot easily be imagined, they got on well, and Klein would go out of his way to speak warmly of the older man. Gordan's best result was in the theory of binary forms (homogeneous polynomials in two variables and of any degree). In 1868 he had showed that the ring of invariants and covariants of any binary form is finitely generated as a ring (see Box 2.3). His proof was heavily computational and quite explicit, and no one had been able to make significant progress on the case of 3 variables. Hilbert and Gordan spent a good week together in Leipzig in the spring of 1888, after which Hilbert elatedly reported to Klein: 'With the stimulating help of Prof. Gordan an infinite series of thought vibrations has been generated within me, and in particular, so we believe, I have a wonderfully short and pointed proof for the finiteness of binary systems of forms.'[9] Hilbert had caught fire. A week later, when he met with Klein in Göttingen, he had already put the finishing touches on the first in a series of papers on algebraic invariant theory that established him as a research mathematician of great originality. Hilbert quickly established the finiteness for any system of invariants for forms of arbitrary degree, but by methods that were not constructive. They were general existence theorems.

At this point an aspect of Hilbert's approach to mathematics surfaces clearly for the first time that was going to be the source of many of his later successes and occasion many later controversies. It concerns the way a mathematician can show that something can

---

[9] Hilbert to Klein, 21 March 1888, in Frei (1985) p. 39.

Box 2.3 Invariant theory

The simplest polynomial expression, other than the linear, is the quadratic. In two homogeneous coordinates it appears as $ax^2 + bxy + cy^2$. The corresponding equation $ax^2 + bxy + cy^2 = 0$ for the ratio $x{:}y$ either has two distinct (real or complex) solutions, or a repeated solution, and these cases are distinguished by the value of $b^2 - 4ac$. This follows from the familiar fact that the solutions are given by

$$\frac{x}{y} = \frac{-b \pm \sqrt{b^2 - 4ac}}{2a}.$$

Invariant theory addresses this question by investigating the effect of linear changes in the variables on the given polynomial f. The transformation T given by

$$\tilde{x} = a_{11}x + a_{12}y$$
$$\tilde{y} = a_{21}x + a_{22}y$$

produces a quadratic in $\tilde{x}$ and $\tilde{y}$, and the corresponding expression for $b^2 - 4ac$ turns out to be $(b^2 - 4ac)(a_{11}a_{22} - a_{12}a_{21})^2$. Accordingly, the expression $b^2 - 4ac$ is an invariant, because it changes under the transformation only by a multiple of a power of the determinant of T.

Nineteenth century invariant theory is the study of a general homogeneous polynomial of any degree and in any number of variables. Invariant theorists searched for expressions in the coefficients of the given polynomial and its variables which, when a linear transformation of the variables is made, change only by a multiple of a power of the determinant of the transformation. Such expressions that involve the variables were called covariants; ones in the coefficients alone were called invariants. Mathematicians soon found that the quadratic in two variables is deceptively simple.

Consider for example a cubic in two variables, given by a homogeneous polynomial of the form:

$$f(x, y) := ax^3 + 3bx^2y + 3cxy^2 + dy^3.$$

cont. p. 25

The 3s introduced here simplify the formulae coming up below. The early invariant theorists, Cayley and Sylvester in England, Aronhold, Clebsch and Gordan in Germany, and Hermite and Jordan in France, showed that coordinate transformations applied to the binary cubic revealed the existence of an invariant:

$$a^2d^2 - 3b^2c^2 + 4b^3d + 4ac^3 - 6abcd.$$

They called this invariant the discriminant of the cubic. More precisely, the same linear transformation of the variables, T, that we used before produces a new cubic polynomial, and this invariant is multiplied by the fourth power of the determinant of T. The size of this invariant is an important warning that matters become very complicated very quickly, and even leading mathematicians found the going increasingly difficult. They also satisfied themselves that this is the only invariant of the cubic.

When they considered polynomials of degree 4 they found it had two distinct invariants, one of degree 2 in the coefficients and one of degree 3. All other invariants could be expressed as polynomials in these two. They found that polynomials of degree 5 have four invariants, and every other invariant is a polynomial in these four. They can be labelled by their degree in the coefficients: I_4, I_8, I_{12}, I_{18}, and they have, in their simplest form 12, 59, 228, and 848 terms respectively (listed in books of the period). Remarkably, Sylvester discovered that these four invariants are not independent, but are related by an equation (of degree 36 as an expression in the coefficients) which he called a syzygy.

Covariants do not appear in the invariant theory of binary quadratic forms. In the case of cubic polynomials the simplest is the so-called Hessian, $H = H(f)$, which is this covariant of degree 2 in the variables:

$$f_2 := (ac - b^2)x^2 + (ad - bc)xy + (bd - c^2)y^2.$$

cont. p. 26

There is also this forbidding expression of degree 3:

$$J := (a^2d - 3abc + 2b^3)x3 + 3(abd + b^2c - 2ac^2)x^2y$$
$$- 3(acd - 2b^2d + bc^2)xy^2 - (ad^2 - 3bcd + 2c^3)y^2$$

These expressions are connected by a syzygy of their own: $4H^3 = df^2 - J^2$.

When Hilbert lectured on this material in 1897, he showed how this syzygy controlled the solution of the cubic equation $f(x, y) := ax^3 + 3bx^2y + 3cxy^2 + dy^3 = 0$. To summarise his results:

1. if $d \neq 0$ then the cubic has three distinct roots;
2. if $d = 0$, $H \neq 0$ then the cubic has a double root and a distinct third root;
3. if $d = 0$, $H = 0$ then the cubic has one triple root.

If we pause to take stock we can see that a number of claims are being made about binary forms of various degrees. They have invariants and covariants—even this is not obvious for homogeneous polynomials of large degree. All invariants and covariants of a binary form may be expressed as polynomials in a finite set of invariants and covariants (which form a basis), and there may be polynomial equations (syzygies) connecting these basic invariants and covariants. There is plainly some sort of theory at work that validates these claims. Indeed, Sylvester had a way of enumerating invariants and covariants of a given degree. And with considerable effort Paul Gordan had shown in 1868 that there is always a finite basis for the invariants and covariants of a binary form of any degree. At that point the immense technical difficulties involved in dealing with forms in three or more variables blocked everyone's path.

The first novelty in Hilbert's presentation of the subject as his 14th problem was to admit other groups of transformations. As Hilbert saw it, the theory of the invariants and covariants of a form in n variables when all linear transformations are considered was now well worked out. But the theory

cont. p. 27

for subgroups was much less well understood. The smaller the group, the less its capacity to contain transformations turning one object into another, and so the number of distinct objects will go up. It is by no means clear that the invariants and covariants will retain a finite basis. L. Maurer had claimed, in 1899, that it was still the case that there was a finite basis, but his proof turned out to be flawed.

The way forward was first to consider the most important groups in mathematics, and then to strive for generality. The important groups, in this context, are the Lie groups and the group of permutations on n symbols. Then one might ask about all groups that can be faithfully represented as groups of matrices (the so-called linear groups).

be done, and a problem solved. The obvious way, the mainstream way, is to produce an answer explicitly. If the question asks for the solution of the equation $x^2 = 4$, then the process of taking square roots produces the answers $x = 2$ and $x = -2$. For more complicated problems, more complicated arguments are needed. Paul Gordan, in particular, was a master at manipulating long algebraic expressions; some of his equations run for several pages. But there are questions which cannot so easily be answered in this way. Euclid, for example, showed that there is an infinite number of prime numbers, but he did so by arguing indirectly.

There are other proofs of the infinitude of prime numbers, but no one has yet been able to seize the prize, a formula which would give the nth prime number as a function of n.

The direct method has one clear advantage over any indirect one: it gives an explicit answer that can be put to other uses. It is hard to imagine science proceeding very far if scientists could not calculate numbers and plug them into other expressions. But there is an art to using it. In well-trodden branches of mathematics the manipulation of symbols is routine, and the good mathematician proceeds quickly and well. Algebra is reliable, one can calculate with it for days at a stretch if need be, and once checked for accuracy nothing

---

Box 2.4 Euclid's proof that there is an infinite number of prime numbers

In Euclid's *Elements*, Book IX Proposition 20, he showed that there cannot be only a finite number of prime numbers. He argued by contradiction. If there were only a finite number, let them be $p_1, p_2, \ldots, p_n$. Then either the number $p_1.p_2.p_n + 1$ is prime, or it is not. If it is prime, there is a contradiction because it is not equal to any of $p_1, p_2, \ldots, p_n$. If it is not prime, then it must be divisible by a prime number, but it is plainly not divisible by any of $p_1, p_2, \ldots, p_n$, so again there is a contradiction. Therefore the assumption that there are only finitely many prime numbers must be wrong, and there is therefore an infinite number of them.

---

more need be done. But when a problem is new, or unexpectedly difficult, symbol manipulation is blind. It is like looking for a faint star with a telescope. No amount of unsuccessful calculation will tell you if tweaking your method will work or if, indeed, there is no solution at all and all your efforts are bound to be in vain.

Indirect methods, which aim to show on other grounds that something must exist, or something cannot exist, have the advantage, when they work, of bringing conceptual clarity. If they show that something must exist they are a powerful stimulus to go back out and find it. But, as Hilbert was to discover, they can have an air of the miraculous about them that some mathematicians find hard to accept.

Hilbert's breakthrough in invariant theory consisted in finding just such a general method. He took the question 'is there a finite basis for these objects?' placed it in a more abstract context, and showed that there must be a basis by little more than using induction on the number of variables involved.[10] The whole proof barely

---

[10] Mathematical induction is a process whereby some claim is proved for every positive integer in these two steps: (1) prove it is true when the integer takes the value 1, (2) prove that if the claim is true when the integer takes the value n then it is true when the integer takes the value $n + 1$.

lasts two pages. As he commented in a letter to Klein when he sent one of his first notes describing his work for publication, 'I have restricted the use of formulae as far as possible, and only presented the intellectual content in a crisp manner.'[11]

Gordan's initial response has become famous, and must have spread rapidly around the seminar rooms and coffee tables where mathematicians liked to gather: 'This is not mathematics, it is theology.'[12] Hilbert himself admitted that his method gave no means of finding the basis, no indication of its size, no impression of what the elements of the basis might look like. The task would be much easier if you could show that the number of variables determined the number of elements in the basis. It would then be enough to find that many elements which were independent of each other, and they would necessarily be a basis for the problem at hand. It would be even better if you had a theorem which showed that in any given problem the basis elements were polynomials of degree no greater than a number you could write down in advance. Then your task would reduce to considering only polynomials of that degree or less. But as matters stood in 1888, no such theorems were to hand.

Gordan later conceded that even theology has its uses. But he raised other objections in a letter to Klein in 1890.[13] He did not dispute the importance of the result, or its validity, but he disputed that the inductive step had been validly carried out with sufficient clarity. Hilbert replied to Klein that what Gordan seemed to want was a different kind of proof, and he would of course be pleased to see one if it was shorter and more compelling than his. But his argument was a proof, he had shown it to many people and listened carefully to their replies and no-one had objected. Klein met Gordan in April for a week and smoothed over the difficulties. But on the other hand, Hilbert felt the shortcomings of his conceptual approach rather keenly, and in the late 1880s he began to remedy the problem, moving the subject on dramatically, and by

---

[11] Hilbert to Klein, 12 December 1888, in Frei (1985) p. 45.

[12] Blumenthal (1935), p. 394.

[13] Quoted in Frei (1985), p. 65.

1897, as he observed in his lectures 'We can nevertheless not be satisfied with merely knowing the number of invariants, as it is even more important to also know about the in- and covariants themselves, and about the relations between them.'[14] He never completely succeeded, but he made significant progress, and after more work by a series of mathematicians between 1900 and 1930 the problem was taken up again with the arrival of high-powered computing and is now pretty well solved.

Hilbert's discoveries had been published as three short notes in a journal of Göttingen University. These were statements of results and, as was quite customary, no proofs were given. The longer papers, complete with proofs, appeared in the more prestigious *Mathematische Annalen*. This was also a Göttingen-based journal, established in 1869 at the instigation of Clebsch (another Königsberg product) and Carl Neumann as a rival to the Berlin-based *Crelle's Journal*. Clebsch died in 1872 at the age of 39, and the journal had eventually passed to Felix Klein. He promoted it energetically, as part of the long-running rivalry between Göttingen and Berlin. He sought articles on a wider variety of topics than were current in Berlin, and from many more nationalities. The Berliners took the understandable view that theirs was the leading university for mathematics in the world at the time, and stood by their judgements as to what were the most important branches of the subject. Geometry was not central to their interests. Klein took the equally understandable view that you could not confront Berlin head-on, but you could outflank it by looking further afield and being open to more novel departures. And geometry was central to his whole vision of mathematics. There is no doubt that he had noted Hilbert's brilliance early on, and set about cultivating his contact with him. Securing Hilbert's papers on invariant theory for the *Annalen* was good for business. Good for the journal, good for Klein, and good for Hilbert. Not for the last time the fortunes of these very different men, Hilbert and Klein, rose together.

Hilbert's papers also opened up something else: a gap between his views and those of Leopold Kronecker on what it means for

---

[14] Hilbert (1993) p. 61.

something to exist. What Gordan had referred to as theology (I suppose because Hilbert's basis elements are no more visible than angels) Kronecker would have regarded as illegitimate. Over the years Hilbert's position hardened, and he became a well-read but hostile exponent of Kronecker's views, which he would set out only to deplore.

Kronecker's notoriously difficult style laid great store by explicit algorithmic procedures.[15] His philosophy of mathematics is usually expressed in negative terms: Kronecker was a strict finitist with no place for transcendental numbers, even, on some views, algebraic numbers. This view, although widespread, is not a fair summary of his position. Kronecker would rather have said that he wanted a common method for dealing with all the problems of mathematics that involve properties of polynomials in any finite number of variables over some field, usually the rational numbers or fields whose elements are quotients of polynomials in several variables. So the subject matter included all of algebraic number theory and algebraic geometry, and they were to be treated as two halves of the same subject. The basic building blocks were two things: the usual integers and the rational numbers, on the one hand, and variables on the other. These were combined according to the usual four laws of arithmetic; root extraction was to be avoided in favour of equations (for example, the variable x and the equation $x^2 - 2 = 0$, rather than $\sqrt{2}$).

His programme drew a strong positive response in some circles.[16] Kronecker said he had been led to it on strictly pragmatic grounds; it was easier to study theoretical problems in algebra by taking an equation and all of its roots rather than by picking one arbitrarily (better to study $+\sqrt{2}$ and $-\sqrt{2}$ rather than just $\sqrt{2}$.) Kronecker did not say, or at any rate did not insist, that algebraic numbers do not exist, but he argued that thinking of them in isolation from how they come about does not help. The productive way forward was to recognise that they arise as the roots of polynomial equations with rational coefficients (and so simultaneously with all the other roots

[15] For a spirited defence, see Edwards (1980), 355,

[16] See Gray (1998/99).

of the same equation.) So when Klein, doubtless echoing Hilbert, said that 'Kronecker, . . . wished to banish the irrational numbers entirely for philosophical reasons and recognised the existence of only the integers or at most the rational numbers', he was wrong about the motives and partisan in his view of what it is for something to exist in mathematics. Kronecker's view of an irrational number (say one such as $\sqrt{2}$ defined as the root of a polynomial equation) was that one could always find intervals sequences of nested intervals shrinking down arbitrarily close to any real root. (In the case of $\sqrt{2}$ the interval (1.4, 1.5), the interval (1.41, 1.42), and so on.) Then, he said, 'The so-called existence of real irrational roots of algebraic equations is based purely and simply on the existence of intervals of this kind'.

Hilbert's view, by contrast, was that algorithmic procedures often get in the way, and clarity is gained by seeking abstract methods, often ones he later couched in axiomatic terms. It may be that his early success with invariant theory inclined him in that direction, but it suited his temperament and was to become his creed. In one of his last papers on invariant theory, which Klein presented on his behalf at the International Mathematical Congress held in Chicago in 1893, Hilbert put his views on the style of mathematics in a passage that has become famous. 'In the history of a mathematical theory three distinct periods of development can usually and clearly be distinguished: the naïve, the formal, and the critical.' He located Cayley and Sylvester in the first phase, Clebsch and Gordan in the second, and himself in the third. The results he referred to as 'critical' were all about the abstract existence of invariants; 'critical', it would seem, meant for Hilbert standing back and asking 'what is it we are really trying to do?' and not getting too immersed in the technicalities.

Gordan exemplifies many a specialist mathematician in that he mastered a branch of the subject and stayed there, solving problems and discovering new results for the rest of his career. Hilbert might have been expected to do so. Despite some historians' stories to the contrary there was more than enough to do. But, as he put it at the end of one of his papers, 'Here, I believe, the most important and general tasks confronting the theory of invariants have been resolved.' So he did a shocking thing. He walked away from the subject altogether. He finished his last and most constructive analysis of the problem of exhibiting bases for invariants, and on 29 September, 1892, wrote to his friend Minkowski 'I shall now definitely leave the field of invariants and turn to number theory.'[17] And so he did, pausing only to give some rather elegant courses of lectures on the topic.[18] The effect was surely to underline the feeling that Hilbert was somebody special, who not only does brilliant work but can then abandon the topic, confident presumably that he can do it again somewhere else.

How Hilbert came to Göttingen

The years after 1890 were marked by a changing of the generations in the German mathematical world. Kummer had retired from Berlin in 1883 and been replaced by Fuchs. Fuchs was a former student of Kummer's who had then come under the influence of Weierstrass, and he had distinguished himself as an expert in the theory of differential equations in the complex domain, but he was not of the high calibre of his predecessor. He was a weak, well-meaning man, inclined to seek compromises, and Berlin was in the final stages of a long, acrimonious wrangle between Kronecker and Weierstrass.[19] The dispute was over the implications of Kronecker's philosophy of mathematics, which Weierstrass regarded as fatal to the entire approach to analysis he had built up and made central to the entire mathematical enterprise in Berlin. So although he was in worsening health he hung on to his position as the Head of the Department, fearing what would happen if he retired and Kronecker took over. Then fate intervened, and Kronecker died in 1891. That enabled Weierstrass to retire, and a new successor to be appointed. This turned out to be another of Weierstrass's protegé's, Hermann Amandus Schwarz. He too was a complex analyst, but a certain light is surely shed on him by the fact that he produced the

---

[17] Blumenthal (1935), p. 395

[18] One of these has recently been translated into English, Hilbert (1993).

[19] See Biermann (1988).

first edition of his Collected Works in 1890, when he was 43, which even mathematicians might think was rather early to be concluding one's research career. Berlin was not likely to remain the dominant force in German mathematics with such new men in charge.

Schwarz's appointment was only one of a number of moves that the changes at the top provoked. His move from Göttingen to Berlin freed up a professorship there. Prussian appointments went through the Ministry of Education, and the Minister was Friedrich Althoff, a singularly powerful figure.[20] Klein, who had been at Göttingen since 1886, had been cultivating him for several years, however. They had met during in the Franco-Prussian War, and resumed contact in the 1880s. Althoff was sympathetic to the argument that it was in Prussia's interests to build up a Mathematics Department to rival Berlin's, and he was more than willing to listen to Klein's ideas about how this could be done. But he was not always willing to agree. Klein had proposed Hurwitz and Hilbert in that order, but he was over-ruled, and Heinrich Weber was appointed instead. Hilbert had to content himself with a promotion to an *Extraordinarius* at Königsberg, which came through when Hurwitz took up a full Professorship at the ETH in Zurich. But within a year he too became a full Professor when Lindemann moved to Munich. Now Hilbert could play the nomination game, and he was able to appoint Minkowski as his successor. For the first time in several years they were back in Königsberg.

The academic roundabout continued to turn, and in late 1894 Weber decided to go to Strassburg (then a German University). Klein could now try again, and this time was to succeed. He wrote to Hilbert[21]

> ... You probably do not know yet that Weber is going to
> Strassburg. The faculty will meet this evening, and as little as I can
> know ahead of time what the commission will recommend, I will
> still inform you that I will make every effort to see that no one
> other than you is called here in the first line. You are the man whom
> I need as my scientific extension: for the direction of your work, the

---

[20] See Zassenhaus's essay in Rüdenberg and Zassenhaus (eds.) (1973), pp. 22–26.
[21] Klein to Hilbert, 6 December, 1894, in Frei (1985) p. 115.

power of your mathematical thinking, and the fact that you now stand in the middle of your productive career. I reckon that you will give a new inner strength to the mathematical school here, which has grown and, as it appears, will continue to grow a great deal further, and perhaps you will exert a rejuvenating influence on me as well. These are the ideas that I and my friends here would welcome you with. Indeed, I cannot know whether I will prevail in the faculty, even less so whether the recommendation we make will ultimately be followed in Berlin. But this one thing you must promise me, even today: that you will not decline the call if it comes to you!

In the event, some of Klein's colleagues criticised his choice, presuming that he wanted to appoint an easy-going, amenable younger man. To this Klein replied: 'I want the most difficult of all'.[22] He prevailed. Hilbert came to Göttingen in 1895, and the remarkable rise of Göttingen in the mathematical world could begin.[23] The contrast between Hilbert and Klein was marked, but they made a productive partnership. Both were very hard workers. But Hilbert's colleagues did not think of him as an organiser, nor did they regard him as particularly well read—in those respects he was outstripped by Felix Klein, who really was building up the Göttingen Empire. They thought of Hilbert as 'a man of problems', some one with a tremendous will, who truly believed every problem can be solved.

Hilbert and number theory

The confidence Hilbert had in his own abilities was widely shared. The German Mathematical Society (the DMV[24]) had been founded, after some false starts in the 1870s, in 1890 with Georg Cantor as its first president.[25] Its creation reflected an international move

---

[22] Blumenthal (1935), p. 399.

[23] The rise of Göttingen, and the remarkable combination of Klein and Hilbert, is analysed in Rowe (1989).

[24] The Deutsche mathematiker Vereinigung. Its journal is the *JDMV* or the *Jahrsbericht den Deutschen mathematiker Vereinigung.*

[25] Dauben, (1979), p. 159.

toward the creation of specialist societies and specialist journals, but the great strength of mathematics in Germany gave their Society a particular character. They set about commissioning extensive reports on the state of the art in this or that subject, and in 1893 they asked Hilbert and Minkowski to report on the theory of numbers. The choice of Minkowski was obvious. He had established himself as a world authority while still a student, and had worked energetically on the subject ever since. But Hilbert had little to show for himself except his work on invariant theory and his expressed interest in the new subject.

The first thing he did was to produce simplified proofs of the transcendence of π and e. Then he began to work intensively on the more algebraic side of the subject, leaving the analytic side to Minkowski. He had discussed the topic at length with Hurwitz on their many walks together, and Hurwitz was an expert guide, but even so the topic must have been a challenging one, because it had been a favourite topic of many of the best German mathematicians of the nineteenth century. It was indeed something of a German speciality, more so than invariant theory, and, not incidentally, one more likely to put him at the apex of the German mathematical world.

The German tradition in the subject begins with Gauss, who dominated mathematics for the first half of the nineteenth century. He once described number theory in these terms: 'Mathematics is the queen of the sciences and arithmetic the queen of mathematics.'[26] It was his achievement to define the subject so that it seemed to have deep significance, but to his views must be opposed another: number theory as the Great Pretender, a subject that claims the throne of mathematics while in fact being obscure and irrelevant. Some held that even the supposedly most important problems in the subject were not that interesting, while others held that while there were some significant problems the vast bulk of the subject consisted of mere puzzles leading nowhere.

The historian of mathematics can at least observe that the subject has not always been highly regarded. Fermat himself believed

---

[26] Waltershausen (1856), p. 79.

that his efforts to interest his contemporaries had failed. Towards the end of his life he wrote to Huygens 'Maybe posterity will be grateful to me for having shown that the ancients did not know everything.'[27] He can have had no confidence that what was by then known as Fermat's Last Theorem would become a favourite problem among nineteenth century mathematicians, still less that any one would, in Hilbert's words in Paris, be 'led by their interest in this problem to formulate a theorem that stands at the centre of the theory of numbers and whose significance extends far beyond the boundaries of number theory, into the realm of algebra and the theory of functions.'

Even these few words of Hilbert's say more than is at first apparent. It is probably not true that Kummer was inspired by the lure of Fermat's Last Theorem. The historian of mathematics H.M. Edwards has argued that in fact Kummer's first interest was in generalising Gauss's work on quadratic reciprocity to questions about cubes and higher powers, and only later, when he saw a tempting connection to Fermat's Last Theorem, did he take his work in that direction.[28] This is not an idle quibble, because the name that is missing from Hilbert's little speech is that of Gauss, who was generally disdainful of Fermat's Last Theorem. As he wrote to Olbers in 1816: 'Indeed I regard Fermat's [Last] Theorem as an isolated result for which I have little interest, for it is easy to think of collection, of such results which one can neither prove nor contradict . . . . I would only be persuaded if . . . . some break-through in the theory [of cyclotomy] allowed Fermat's Theorem to appear as one of the less interesting corollaries.'[29]

In fact, Hilbert was establishing a pedigree for the subject of number theory, by arguing that a line of distinguished mathe-maticians have worked on the subject, and that the ultimate reason for its importance is that it reaches beyond itself to embrace other domains of mathematics, whose importance cannot be doubted. To that end he was even willing to allow that problems like Fermat's

---

[27] Quoted in Weil (1984), pp. 118–119.
[28] See Edwards (1975), p. 226.
[29] Gauss, *Werke*, 10(1), pp. 75–76.

Last Theorem can seem unimportant. In fact, Hilbert would have agreed that some problems in number theory actually are unimportant, but those that reach outwards far enough are the ones that matter. This interplay between concrete problems and big, general ideas was indeed what Hilbert was able to articulate so well, in his own work and in the problems he presented in Paris.

It is also clear that Hilbert knew very well he was taking a German subject to a French audience that, although he was too polite to say so, had not appreciated it fully. From Gauss to Kummer and Kronecker, from Gauss to Dirichlet and then Dedekind, and thence to Hilbert and Minkowski, algebraic number theory had been largely German, largely indeed Prussian. It had blossomed better in a university culture where mathematics was to be done for its own sake in the confident expectation that applications would flow naturally, than in the Parisian environment that placed the connections between mathematics and physics at the centre of the enterprise. Catherine Goldstein, another historian of mathematics, has analysed the French situation in some detail, and she finds that there were several flourishing groups of number theorists in France, but that most of them disparaged the use of advanced mathematics. They sought, and sometimes found, elementary solutions to problems that were, in their view elementary because they were only about the integers. At its best this work was explicit and effective, less abstract than Hilbert's but better at such non-trivial topics as finding prime factorisations. Problems with their work arose when the problems they chose were precisely those that could be solved in an elementary way. This form of self-denying ordinance was antithetical to Hilbert, who also had a formidable command of the modern methods that the French groups had quite often never mastered. The view that large amounts of Galois theory were needed to tackle the important problems in number theory was not universally agreed. It was, as Hilbert well knew, the view he was trying to persuade others to accept.

The German tradition took certain themes which can be found in Gauss's monumental *Disquisitiones Arithmeticae* of 1801, rearranged them, and extended them. The problems are not that difficult to state. For example, can you solve this equation in integers: $x^2 - 15y^2 = 2$? The answer is no, because if you could, then suppose a solution is $x = m$, $y = n$. Then we have $m^2 - 15n^2 = 2$. If we divide both sides by 5 and take just the remainders, it follows that there is a number, m, with the property that the remainder on dividing m^2 by 5 is 2. But it is easy to see that there is no such number. The only possible remainders of a square on division by 5 are 0, 1 and 4. One of the first things Gauss did in his *Disquisitiones Arithmeticae* was to explain in careful detail how remainder arithmetic works and to put it on a sound footing as part of arithmetic. It is nowadays taught to children. Mathematicians know it as modular arithmetic, and speak of working modulo a number (usually a prime); the example above led us to the problem 'is 2 a square modulo 5?'.

Gauss went on to consider such questions as: what numbers can be written in the form $x^2 - 15y^2$? This is an old question that goes back as far as Fermat, who showed that the numbers which can be written as a sum of squares (that is, in the form $x^2 + y^2$) are, 0, 1, 2, all squares, prime numbers of the form $4n + 1$ (thus $5 = 2^2 + 1^2$, $13 = 2^2 + 3^2$, and so on) and numbers whose prime factors are of this form. The general topic of what numbers can be expressed by a given quadratic form (that is, written in the form $ax^2 + bxy + cy^2$) was tackled by both of the leading mathematicians of the eighteenth Century: Euler and Lagrange. Gauss's most original contribution was to show that these expressions can be manipulated to produce others and so illuminate what is going on. His way of combining two expressions to produce a third of the same kind would eventually be seen as bringing structure to the subject and with it the kind of conceptual understanding that mathematicians tend to like.

On the way, Gauss solved a famous question. In our simple problem 'is 2 a square modulo 5?' we could use elementary methods. But a good theoretical analysis of the problem requires that one can answer such questions as 'is m a square modulo a prime p?' for very large numbers m and p. What was noticed by Euler, and then addressed directly by Legendre, a leading French mathematician during the period of the French revolution and afterwards, was that when m is a prime number, say q, there is an intimate relation between the question 'is q a square modulo a prime p?' and its 'reciprocal': 'is p a square modulo a prime q?' This is a striking fact because on the face of it it is unexpected. Legendre gave a purported proof, but as Gauss saw, it was deeply flawed, and his own proof in the *Disquisitiones Arithmeticae* is actually the first.

Gauss's proof is difficult, and has attracted harsh criticisms for its obscurity over the years. He went on to give five more proofs, and today a great many are known (a figure of over a hundred is bandied about). Gauss's own were devised not just to replace the original proof with a simpler one but to make connections with other parts of the subject and to get beyond squares to questions about equations of degrees 3, 4 (and beyond, although Gauss stopped at 4). This profusion of work is eloquent. It makes it clear that the topic was suddenly interesting, and that what had drawn the interest was the surprising difficulty in providing proofs, and the unexpected connections between topics that the search for proofs was likely to reveal.

Another topic that Gauss introduced, although he did not regard as lying at the heart of number theory, has the obscure name of cyclotomy. Later generations disagreed, and it rapidly became integrated with the subject. It concerns numbers which are the nth roots of unity, such as $i = \sqrt{-1}$, which is a 4th root of unity (because $i^4 = 1$). Gauss had been led to them in pursuit of his first piece of original work, when he showed that a regular 17-sided figure is constructible by ruler and compass alone.

Constructing the regular pentagon by ruler and compass was research mathematics for the ancient Greeks[30] but by 1797 it was old news. However, no other regular figures with a prime number of sides had been constructed (except for the triangle, of course) and the topic had died of inanition. By relocating it in algebra Gauss saw that a regular n-sided figure would be constructible if and only if the corresponding nth root could be found by solving a sequence of quadratic equations. In particular, this was the case for

---

[30] Although it was old news when it appeared in Euclid's *Elements* Book IV, Prop. 11.

Box 2.5a Quadratic Reciprocity

The theorem of quadratic reciprocity answers the question: when is a number a square modulo a prime p, (the useful technical term is a *quadratic residue*) and when is it a *quadratic non-residue* (and not a square modulo p).

Legendre presented the law of quadratic reciprocity in a most succinct form by means of what is today known as the Legendre symbol, which is defined as follows for prime numbers p and q,

$$\left(\frac{p}{q} \right) = \begin{cases} 1 \ \textit{if } p \textit{ is a quadratic residue mod } q \\ -1 \ \textit{if not.} \end{cases}$$

(it can be extended to pairs of numbers with no common factors).

The theorem of quadratic reciprocity, is the striking claim that, for odd primes p and q:

If at least one of p, q is $\equiv +1 \pmod 4$, then $\left(\dfrac{p}{q} \right) = \left(\dfrac{q}{p} \right),$

and if, however, $p \equiv q \equiv -1 \pmod 4$, then $\left(\dfrac{p}{q} \right) = -\left(\dfrac{q}{p} \right).$

In the first case, p and q are either both quadratic residues or else non-residues with respect to the other, whereas in the second, one of the two is a quadratic residue and the other a non-residue. These relationships may be reformulated even more succinctly as follows:

$$\left(\frac{p}{q} \right) = (-1^{\frac{p-1}{2}\frac{q-1}{2}}) \left(\frac{q}{p} \right).$$

The prime 2 can also be dealt with in this theory, since

$$\left(\frac{2}{p} \right) = (-1)^{\frac{p^2-1}{8}}$$

> **Box 2.5b Quadratic reciprocity in action**
>
> This theorem makes the process of determining if a number is a quadratic residue modulo another almost trivial. For example, is 105 a quadratic residue modulo the prime number 127? We compute
>
> $$\left(\frac{105}{127}\right) = \left(\frac{3.5.7}{127}\right) = \left(\frac{3}{127}\right)\left(\frac{5}{127}\right)\left(\frac{7}{127}\right)$$
>
> (this is what is meant by the character being multiplicative), so
>
> $$\left(\frac{105}{127}\right) = \left(\frac{3}{127}\right)\left(\frac{5}{127}\right)\left(\frac{7}{127}\right) = -\left(\frac{127}{3}\right)\left(\frac{127}{5}\right)\left(-\left(\frac{127}{7}\right)\right)$$
>
> $$\left(\frac{105}{127}\right) = -\left(\frac{1}{3}\right)\left(\frac{2}{5}\right)\left(-\left(\frac{1}{7}\right)\right) = -1.-1.-1 = -1$$
>
> so the answer is no, 105 is not a quadratic residue modulo 127. Notice that this has not involved computing any large numbers, and has avoided searching all the squares of numbers modulo 127.

$n = 17$, and generally for prime numbers of the form $2^{2^k} + 1$. This was the first new result in the topic since the days of Euclid, and Gauss gave full details of this method in the *Disquisitiones Arithmeticae*.

Cyclotomic numbers was another topic that caught the attention of later number theorists. In 1847 there was something of a scandal concerning them, when a number of French mathematicians thought they could be used to solve Fermat's Last Theorem.[31] It is possible to give a natural definition of when a cyclotomic number should be called a cyclotomic integer, and to say when a cyclotomic integer is a prime cyclotomic integer. The tempting thing to do was to seek to show that Fermat's Last Theorem, for some exponent n greater than 2, did not have a solution in cyclotomic integers, and then deduce that no solution in ordinary integers could exist.

---

[31] See Edwards (1977), Chapter 4.

Box 2.6 The regular pentagon

Consider the algebraic number $e^{\pi i/5} = \xi$. It and its powers satisfy the equation $x^5 - 1 = 0$. Plotted in the plane of complex numbers they form the vertices of a regular pentagon.

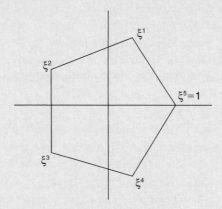

Now, ξ is in fact a solution of the equation

$$x^4 + x^3 + x^2 + x + 1 = 0 \qquad (*)$$

because it is easy to check that

$$(x - 1)(x^4 + x^3 + x^2 + x + 1) = x^5 - 1 = 0,$$

and certainly $\xi \neq 1$. Consider now the equation

$$x^2 + x + 1 + x^{-1} + x^{-2} = 0 \qquad (**)$$

obtained by dividing equation (*) by x^2. Set $x + x^{-1} = y$, so $(x + x^{-1})^2 = x^2 + 2 + x^{-2} = y^2$ and equation (**) becomes

$$y^2 + y - 1 = 0.$$

This is a quadratic equation for y, so it can be solved geometrically by ruler and compass alone, take the solution which corresponds to, $y = (-1 + \sqrt{5})/2$. Now solve the equation $x + x^{-1} = y$ in the form $x^2 - yx + 1 = 0$, which you can do because it is another quadratic equation and you know y. You have now constructed the quantity $x = \xi$ geometrically by ruler and compass, which means that the regular pentagon is constructible in that fashion.

Gabriel Lamé informed the Academy of Sciences in Paris that he had discovered such a proof, but in the course of it he manipulated the new integers as if they obeyed all the rules that genuine integers obey. In particular, he assumed that a cyclotomic integer factors into prime cyclotomic integers in an essentially unique way.

Joseph Liouville, who edited a journal and made it his business to keep in touch with developments in Germany, sounded a note of caution at this point, but even so the forceful figure of Cauchy joined in on Lamé's side (claiming to have a similar proof of his own, which Lamé may not have thought was exactly on his side). Liouville then wrote to Kummer, who was then a young number theorist inspired by Jacobi's work on a generalisation of the quadratic reciprocity theorem and in possession of a fully fledged theory of these unusual 'integers'. It emerged that he had already discovered that certain cyclotomic integers can be factored in several distinct ways. This destroyed the simple proofs that Lamé and Cauchy were hoping for, although it kept Kummer working on Jacobi's questions, and was to lead to a significant (but much more complicated) breakthrough in the study of Fermat's Last Theorem.

Gauss had therefore set out a range of topics that a century of mathematicians were to embrace: quadratic reciprocity and its generalization to higher powers, quadratic forms, and cyclotomy. As he had predicted on the basis of his own work there were a number of deep connections between these topics and between them and other branches of mathematics, that made proofs of even elementary statements in the subject difficult to find, and if you had a taste for such matters such proofs were also particularly satisfying. The result was a huge theory, the cumulative production of such mathematicians as Eisenstein, Jacobi, Dirichlet, Dedekind, Kummer, and Kronecker, that threatened to swallow up the humble integers and their curious properties (after all, what primes are there of the form $2^{2^n} + 1$ for some n?). It was also a theory with conflicting foundations: the viewpoint that Kronecker held had provoked Dedekind into a strikingly different one. It was a natural task for the German Mathematical Society to ask Hilbert and Minkowski to bring it all up to date.

Minkowski, however, soon withdrew from the project because

he felt he was too busy with other things, especially his book *The Geometry of Numbers*. The *Report*, for all its 371 pages, is only a torso. As a result a number of other topics in number theory with a pedigree going back to Euler and Jacobi are under-represented, and this may well have skewed the subsequent impact of Hilbert's problems on number theory. But Minkowski kept in touch with Hilbert, and performed the useful service of checking all the page proofs. His opinion of the *Report* was high. He wrote to Hilbert: 'I congratulate you now that finally the time has come, after the many years of labour, when your report will become the common property of all mathematicians, and I do not doubt that in the near future you yourself will be counted among the great classical figures of number theory'.[32]

He was correct in this prediction, and one reason is the lucidity of Hilbert's style. Hermann Weyl, whose own masterful control of the German language was probably unmatched by any other mathematician, regarded the preface of Hilbert's *Report* as one of the great works of prose in the mathematical literature, and described how as a young man he had heard 'the sweet flute of the Pied Piper that Hilbert was, seducing so many rats to follow him into the deep river of mathematics.'[33] Another is the skill with which Hilbert re-presented the history of the number theory in a way that appealed to its practitioners, making it clear how it had emerged as a systematic science only during the past hundred years or so.

Hilbert's principal contribution was nonetheless that he re-wrote the subject so that knowledge of its history and familiarity with the older texts was no longer required. New readers could start here. He did this in a variety of ways. One was his whole-hearted introduction of the ideas of Galois theory, which formed the second off the five parts of the *Report*. Galois theory was invented to explain when a polynomial equation is solvable by radicals (see Box 2.7). Hilbert thought of it as a theory of algebraic numbers (numbers which arise as the roots of polynomial

---

[32] Rüdenberg and Zassenhaus (1973) p. 100.

[33] Weyl (1944), in *Gesammelte Abhandlungen*, 4, p. 134.

equations) and therefore as an organising principle of (algebraic) number theory. This not only gave a clearer conceptual shape to the subject, it explained why questions in number theory led naturally to and from questions about cyclotomy. It was, indeed, a paradigm example of the benefits of digging deep in pursuit of an explanation, and it put the seal on a century of stretching the definition of an integer until the general concept of an algebraic integer was paramount, and the ordinary integers had to travel with a qualifying adjective (the 'rational integers', a rather awkward phrase).

Hilbert was particularly fond of analogies, which he saw as attempts to unify the disparate fields of mathematics. The analogies he offered were never fanciful, but were intended to suggest deep, if at times obscure, connections. The implication always was that one should seek for the common object in virtue of which two fields were analogous. He noted that a number of problems in number theory relied on complex function theory, and that there were numerous points in which the theory of algebraic functions and fields of algebraic numbers overlapped. (This was the central thrust of the paper by Dedekind and Weber.) He then wrote:

> Thus we see how far arithmetic, the Queen of mathematics, has conquered broad areas of algebra and function theory and had become their leader. The reason that this did not happen sooner and has not yet developed more extensively seems to me to lie in this, that number theory has only in recent years become known in its maturity . . . . Nowadays the erratic progress characteristic of the earliest stages of development of a subject has been replaced by steady and continuous progress through the systematic construction of the theory of algebraic number fields. The conclusion, if I am not mistaken, is that above all the modern development of pure mathematics takes place under the banner of number: the definitions given by Dedekind and Kronecker of the concept of number lead to an arithmetisation of function theory and serve to realise the principle that, even in function theory, a fact can be regarded as proven only when in the last instance it has been reduced to relations between rational integers.

Box 2.7 Solvability by radicals

A polynomial equation is said to be solvable by radicals if and only if its solutions can be expressed in terms of the co-efficients by a formula that only involves the basic operations of arithmetic together with taking arbitrary roots (also called radicals, whence the name).

Thus the quadratic equation $ax^2 + bx + c = 0$ has the two solutions

$$x = \frac{-b \pm \sqrt{b^2 - 4ac}}{2a}$$

and is solvable by radicals. It was shown in the early sixteenth century that the cubic and quartic equations are solvable by radicals, but no one was able to solve the general equation of degree 5 in that way. Eventually it was shown by Abel that this is in fact impossible. Galois then set about creating a theory which explained which equations (of degrees 5 or more) were solvable in this way, and which were not.

Arithmetisation was a vogue word of the day. It had been used not many years before by Klein to capture the reduction of the fundamental concepts of analysis to those of arithmetic. Hilbert added 'The arithmetic of geometry fulfils itself in the modern study of non-Euclidean geometry, which it provides with a rigorous logical structure and gives the most direct possible and completely objection-free introduction of number into geometry'.

Hilbert's breakthrough with invariant theory had been achieved by stepping back from the computational side and taking a more conceptual approach. Not surprisingly, that was his approach to number theory too:

> ... the fifth part develops the theory of those fields [the cyclotomic fields] which Kummer took as a basis for his researches into higher reciprocity laws and which on this account I have named after him. It is clear that the theory of these Kummer fields represents the highest peak attained today on the mountain of our knowledge of arithmetic; from it we look out on the wide panorama of the whole

explored domain since almost all essential ideas and concepts of field theory, at least in a special setting, find an application in the proof of the higher reciprocity laws. I have tried to avoid Kummer's elaborate computational machinery, so that here, too, Riemann's principle may be realised and the proofs carried out not by calculations but purely by thought.

Hilbert and geometry

After less than a decade dominated by the writing of the *Report*, Hilbert repeated his gambit of 1892, and abandoned number theory. The new subject that claimed his attention was something of a shock to his Göttingen colleagues: the foundation of elementary geometry. What they did not know was that Hilbert had already been drawn to the subject in his years at Königsberg. He had not at first been attracted to the teaching of projective geometry, a staple subject he felt he did not have much to say about. But he soon enough found that there were things to say. The subject existed in two forms. One was heavily algebraic, and led towards the study of curves defined by polynomial equations. The other was known as synthetic geometry, and it more closely resembled the arguments found in Euclid's *Elements*. In classical Euclidean geometry one moves line segments and figures such as triangles and circles around *en bloc*, and though the whole subject can be translated into coordinate geometry it is actually very different in flavour. Synthetic projective geometry also moves figures around en bloc, but with a wider class of transformations (figures may be replaced by their shadows under projection from a point source of light). The nineteenth century view, held most articulately by Klein, was that any geometry consisted of a space and a group of transformations, and that projective geometry, having the largest group, was the most fundamental. Or, to put the same point in a different way, projective geometry was the most fundamental because it rested on the fewest initial assumptions. The other geometries were obtained by adding assumptions. This imposed restrictions on the transformations, so the larger the number of assumptions the smaller the group of transformations.

Hilbert cannot have been too excited by the audience his first lectures on projective geometry attracted at Königsberg: two students plus the director of the Royal Art School.[34] But he soon discovered that the experts disagreed about a number of things in synthetic projective geometry. In particular, they disagreed about which theorems followed from which others. In September 1891 he heard a lecture by Hermann Wiener in which he raised the possibility of developing projective geometry axiomatically from Pascal's and Desargues' theorems. This was the occasion for Hilbert to say 'One should always be able to say, instead of 'points, lines, and planes', 'tables, chairs, and beer mugs'.[35] In other words, one should be able to purge one's arguments of words that might smuggle in meanings and imply a validity the logical structure of the argument cannot actually carry. The point was not original with Hilbert, but he grasped its significance more profoundly.

In 1894 Hilbert lectured again on geometry, but it was a letter from Friedrich Schur to Klein which was passed on to him in early January 1898 that really set him off. In the letter Schur raised the possibility of deriving Pappus's theorem without using the Archimedean axiom.[36] This opened the way for a non-Archimedean geometry. It was beginning to look as if there was new life in the old geometry after all. He therefore lectured on the subject at Göttingen in the winter Semester, much to the surprise of Blumenthal, who wrote 'This caused astonishment among the students, for even we older people, participants in the stroll through the number fields, had never known Hilbert to concern himself with geometrical questions. Astonishment and wonder were caused when the lecture began and a completely novel content emerged.'[37]

The surprise was repeated on an international stage when Hilbert published his ideas, in a revised form, as part of a book

---

[34] Hilbert to Klein, 30 June 1891, quoted in Toepell (1985), p. 335.

[35] Blumenthal (1935), p. 403.

[36] See Toepell (1986), p. 338.

[37] Blumenthal (1935), p. 402.

Box 2.8 The theorems of Desargues, Pappus, and Pascal

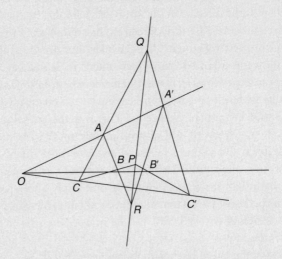

Desargues' theorem asserts that if two triangles *ABC* and *A'B'C'* are in perspective from a point *O* and the lines *AB* and *A'B'* meet at *R*, the lines *BC* and *B'C'* meet at *P*, and the lines *CA* and *C'A'* meet at *Q*, then the points *P*, *Q*, and *R* lie on a line.

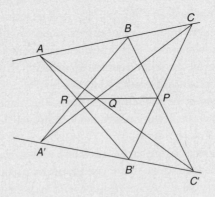

Pappus' theorem asserts that if the points *ABC* lie on one line and the points *A'B'C'* lie on another, and the lines *AB'* and

$A'B$ meet at R, the lines BC' and $B'C$ meet at P, and the lines CA' and $C'A$ meet at Q, then the points P, Q, and R lie on a line.

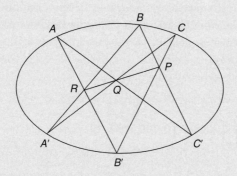

Pascal's theorem asserts that if the points $ABCA'B'C'$ lie on a conic, and the lines AB' and $A'B$ meet at R, the lines BC' and $B'C$ meet at P, and the lines CA' and $C'A$ meet at Q, then the points P, Q, and R lie on a line.

published to mark the unveiling of the statue to Gauss and Wilhelm Weber in Göttingen. The *Grundlagen der Geometrie* (*Foundations of Geometry*) was published in 1899. It ran to ten editions, seven in Hilbert's lifetime, and was translated into several languages. It came to epitomise for many what Hilbert's ideas about mathematics were, and was perhaps more easily understood because it presented them in an elementary arena. Nonetheless, as we shall, see, it carried a number of interesting flaws to match its insights.

It begins with the beer mugs. 'We think of three different systems of things: the things in the first system we call *points* and denote A, B, C, . . .; the things of the second system we call *lines* and we denote a, b, c, . . . ; the things of the third system we call *planes* and denote a, β, γ, ... There then followed five kinds of axiom which determine what one can say about these things. There are axioms of incidence (which enable you to say 'this point lies on this line'),

Box 2.9 Non-Archimedean quantities

The Archimedean axiom asserts that given any quantities x and y with $x < y$ there is a number n such that $nx > y$. In other words, if x is copied often enough the result is quantity bigger than y. The simplest example of how this need not be true is with 'numbers' of the form $a + bt$, with the rule that $a + bt < c + dt$ if and only if either $a < c$ or $a = c$ and $b < d$. We add such 'numbers' according to the rule

$$(a + bt) + (c + dt) = (a + c) + (b + d)t.$$

We multiply them according to the rule

$$(a + bt).(c + dt) = (ac) + (bc + ad)t$$

Then $t = 0 + 1.t$ is less than $1 = 1 + 0t$, but there is no 'number' n such that $nt > 1$. To see this, suppose that $n = a + bt$, then $nt = (a + bt).t = at$, but $at < 1$.

Such 'numbers' are not entirely bizarre. They can be thought of as information about the graph of a function. The 'a' part of $a + bt$ is the value of the function at a specific point, the 'b' part gives the slope of the graph at that point. The rule is that at the specified point this function is larger than that one if either its value is greater or the values are the same but the slope is greater.

order ('this point lies between these two'), congruence, parallels, and continuity (which is where Hilbert located the Archimedean axiom).

At each stage Hilbert was concerned to show what theorems can now be deduced, from which it is clear which axioms are actually required, and to show that the axioms are mutually independent and consistent. He was not entirely successful at establishing independence, but that is a minor flaw. He did show, for example, that Pappus's theorem is independent of Desargues' theorem, and the admission or exclusion of the Archimedean axiom has a decisive effect on the validity of Pappus's theorem. A lot of the work in the book is done by what he called segment arithmetic. He based it on what he confusingly called Pascal's theorem and is actually the result better known as Pappus's theorem. This gave him ways analogous to addition and multiplication of combining segments to get others. It followed that if the geometry was coordinatised, the coordinates would have to obey certain rules. In fact, he showed that when Pappus's theorem is true the coordinates must lie in a system where multiplication is commutative, but Desargues' theorem can be true in a system where multiplication is not commutative.[38] Hilbert also gave a convoluted example of a plane geometry in which Desargues' theorem was false.

Further axioms about congruence gave Hilbert a theory of plane area without recourse to the calculus. This was to inspire the 3rd of his Problems. He showed how to construct geometries in which the Archimedean axiom is false, and nodded at the Italian mathematician Giuseppe Veronese's priority in this regard. He concluded the book, apart from one significant point, with a discussion of an old topic from his new view point: what geometric problems are solvable by ruler and compass. He knew very well that Galois theory solved the question completely from an algebraic point of view. Here he showed that if you are given a single interval and a ruler, then the coordinates of the constructible points are rational expressions in numbers which use the operation of extracting the square roots of sums of two squares a finite number of times. This becomes clearer by example. The rational numbers are allowed. If you have constructed a number ω then by constructing a right-angled triangle with lengths of 1 and ω you can construct the number $\left|\sqrt{(1 + \omega^2)}\right|$ (where the absolute value is meant) . If you start with the real numbers and proceed in this way you obtain the plane of complex numbers, and the coordinates describe Euclidean geometry. But if you start with a single segment you obtain an strange extension of the rational numbers, which Hilbert showed is very impoverished. In particular, you cannot construct a

---

[38] In fact, as Hilbert nearly showed, Pappus's theorem implies Desargues' theorem.

Box 2.10 Descartes and Hilbert on multiplication of segments

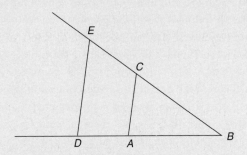

Descartes wrote: For example, let *AB* be taken as unity, and let it be required to multiply *BD* by *BC*. I have only to join the points *A* and *C*, and draw *DE* parallel to *CA*; then *BE* is the product of *BD* and *BC*.

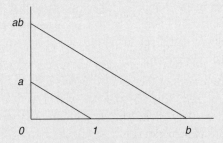

Hilbert wrote: In order to define geometrically the product of a segment *a* by a segment *b* the following construction will be used: Choose any segment which remains fixed during the entire discussion and denote it by 1. Now lay off the segments 1 and *b* from the vertex 0 on a side of a right triangle. Then lay off the segment *a* on the other side. Join the end points of the segments 1 and *a* with a line and through the end point of the segment *b* draw a parallel to this line. It will delineate a

cont. p. 55

segment c on the other side. This segment is then called *the product* of the segment a by the segment b and is denoted by

$$c = ab.$$

Hilbert went on to show that multiplication is commutative: $ab = ba$.

right-angled triangle with a hypotenuse of 1 and a leg of $\left|\sqrt{2}\right| - 1$.[39] So construction with ruler and a given segment is very different from construction with a compass, when it is indeed easy to construct the triangle. Thinking further about this led Hilbert to theorems on fields of numbers in which every positive number is a sum of squares, a topic that was to surface as part of his 17th Problem. He had already studied these in an algebraic context, having first heard of the problem when Minkowski had claimed in his doctoral thesis that it was unlikely that any polynomial in several variables which was never negative was expressible as a sum of squares.[40]

The book ended by observing that negative results in mathematics could profitably be seen as invitations to create something new, and by directing the readers' attention to the purity of its methods, as exemplified by the clarity with which the various axioms played their roles. The clear implication was that other branches of mathematics could profitably be treated in the same way, and in Paris it was to become clear that Hilbert had in mind treating branches of physics in exactly that way.

---

[39] Hilbert showed that if you can construct a number $a + ib$ (where $i = \sqrt{-1}$) then you can construct its so-called conjugate $a - ib$. But in the triangle in question, the other leg would be $\sqrt{2\left|\sqrt{2}\right| - 2}$, and its conjugate would be $\sqrt{-2\left|\sqrt{2}\right| - 2}$, which is imaginary.

[40] Hilbert (1888).

The road to Paris

The annual meeting of the DMV was held in Munich from 17 to 23 September 1899, jointly with the German Scientists and Doctors[41], and Hilbert was among the 80 participants who braved the torrential downpour that marked the start of the meeting. They discussed plans to host the 1904 ICM in Germany. The official invitation would be made, and accepted, in Paris. They reviewed plans to publish more survey articles in the *JDMV*, and they heard numerous lectures. Among these were two by Hilbert. One was on the Dirichlet problem, and the other on the concept of number.[42] Both were themes that were to appear again in Paris.

The most exciting session of the meeting was Ludwig Boltzmann's lecture to the scientists on recent developments in physics.[43] Boltzmann was one of the architects of the theory of thermodynamics, and he was also a brilliant lecturer. He was accustomed to lecture without notes, and the physicist Lise Meitner who attended his lectures for four years said of them: 'After each lecture it seemed to us as if we had been introduced to a new and wonderful world, such was the enthusiasm he put into what he taught.'[44] He took the opportunity to review current controversies in theoretical physics, including the question of whether atoms really existed, and he urged his listeners to throw themselves into the new century. Doubtless they streamed from the lecture theatre quite inspired, but Boltzmann gave them nothing to focus their energies upon. Hilbert, whose rhetorical skills were less dramatic, less witty and erudite than Boltzmann's but no less clear, may have drawn the lesson that substance lasts longer than exhortation, and looking forward does better than looking back.

It was only natural that Hilbert be invited to give a plenary lecture to the 2nd International Congress of Mathematicians, scheduled for Paris in 1900. He consulted with his friends

---

[41] Deutscher Naturforscher und Ärtze.

[42] Hilbert (1904) and (1900), respectively.

[43] Boltzmann, (1900).

[44] Quoted in Cernignani (1998) p. 38.

Box 2.11 The Dirichlet principle

The Dirichlet principle asserts that given any closed curve in space there is a surface of least energy that spans it because any spanning surface will take up a configuration that requires the least energy.

This principle, which is highly plausible on physical grounds, was known to mathematicians from 1870 onwards to be false if the curve is sufficiently crinkly.

Minkowski and Hurwitz, and wondered if he should not reply to Poincaré's address at the first International Congress of Mathematicians (ICM), the one in Zürich in 1897. But Minkowski advised him to seize the moment. He wrote

> Most alluring would be the attempt to look into the future, in other words, a characterisation of the problems to which the mathematicians should turn in the future. With this, you might conceivably have people talking about your speech even decades from now. Of course, prophecy is indeed a difficult thing.[45]

He counselled against themes of a more philosophical nature 'which are probably better for a German audience', and suggested that a technical talk of the kind Hurwitz had given at the ICM in Zürich in 1897 might go over better than mere pleasantries of the kind Poincaré served up, which had not impressed him as much as Boltzmann's lecture.

Hilbert must have worked energetically on his talk. Somewhere along the line he also seems to have decided to turn his own

---

[45] Rüdenberg and Zassenhaus (1973) p. 119, Minkowski to Hilbert, 5 January 1900.

invincible optimism and his confidence in his own energy and ability into a creed for every mathematician. He opposed it to the fashionable pessimism of the eminent physiologist Emil du Bois Reymond[46] and others who foresaw limits to what the science could discover. Du Bois Reymond's famous lecture, given in 1876 and thereafter often reprinted, considered the seemingly invincible march of science only to conclude with the words 'But concerning the riddle of what matter and force might be, and how they can be thought about, the scientist must concur once and for all with the much harder verdict that is delivered: "Ignorabimus".' In mathematics, and even in physics, Hilbert argued, there was no Ignorabimus, no 'we shall not know'. He decided against covering the entire field of mathematics, but to stick closer to what he knew well. It is striking how many of the papers he was to cite had been published in Göttingen journals and had even been presented by him.[47] This can be viewed as evidence of how important Göttingen was in the world of mathematics, or simply as indicative of Hilbert's limited interest in reading the literature. Like many a speaker, he did a poor job of keeping in touch with the organisers and conforming to their schedules, and in June the programme for the meeting was mailed to those who had indicated interest without Hilbert's name being listed. Minkowski was very disappointed, and wrote to Hilbert that he would probably not now go to Paris. Then, in late July Minkowski received a draft of the proposed address. After reading three pages he wrote back striking a note of caution.

> It is original, certainly, to proclaim that providing axioms for arithmetic is a problem for the future, when mathematicians would think that they had possessed these for the longest time. What would lay people in the audience think? Would their respect for us grow? And you will have a fight with the philosophers on your hands.[48]

---

[46] Reid (1970), p. 72.

[47] At the Göttingen Scientific Society members presented papers on behalf of others.

[48] Rüdenberg and Zassenhaus (1973) p. 129, Minkowski to Hilbert, 17 July 1900.

But when he had finished he was more optimistic:

> I can only wish you luck on your speech; it will certainly be the event of the congress and its success will be very lasting. For I believe that this speech, which probably every mathematician without exception will read, will cause your powers of attraction on young mathematicians to grow still more, if that is even possible . . . Now you have really wrapped up the mathematics for the twentieth century and in most quarters you will gladly be acknowledged as its general director.[49]

Prophetic words, indeed.

The Paris ICM is generally agreed to have been something of a shambles. Scott dryly remarked: 'The arrangements excited a good deal of criticism.'[50] It was one Congress of some 200 scientific conferences held in Paris that year in connection with the World Exhibition, as Paris attempted to put itself at the head of the new century. The philosophers met immediately before, for example. The heat—it was August—the crowds, the parallel sessions, the lack of opportunities to meet and talk mathematics, all conspired to detract from the occasion. In retrospect the Congress heard some significant papers on complex function theory and on the history of mathematics. It has been best remembered, however, for only one thing: Hilbert's Problems.

---

[49] Rüdenberg and Zassenhaus (1973) pp. 129–130, Minkowski to Hilbert, 28 July 1900:

[50] Scott (1900), p. 75.

CHAPTER 3

The beacons are lit

When the audience left the auditorium in the Sorbonne they could begin to give their full attention to the printed paper with all the 23 Problems that Hilbert proposed. They would find that it divided into a small number of overlapping topics illuminated by the themes he had set out in his talk. The more they knew about his career to date, the more sense much of it would make.

The first two Problems were ones he had talked about: the nature of the real number continuum and the proposal to give new axiomatic foundations for arithmetic. In the printed paper Hilbert added two more problems that grew more directly out of his work on providing new foundations for geometry. It is a familiar fact of plane geometry that triangles of equal height have areas proportional to their bases. But the analogous claim for triangular pyramids—that two pyramids of equal height have volumes proportional to their bases—has only been proved using the calculus. The plane problem is solved by a truly elementary argument that requires nothing more than cutting the triangles into pieces and reassembling them, but the problem about volumes seems to need a limiting argument involving (not that Hilbert put it so crudely) infinitely many infinitely small pieces. Hilbert observed that this disparity had already been commented upon by Gauss, and suggested that this was for a good reason: it might be possible to find two tetrahedra with equal bases and equal altitudes which cannot be split up into congruent tetrahedra. If they were found it would follow that the calculus was needed for a theory of volumes, but not of area. This would bear on a theme dear to Hilbert: what methods are actually needed to prove what results.

Hilbert's 4th Problem also derived from his study of elementary geometry, but it was found to be rather vague by later investigators. But he had noticed in his study of his friend Minkowski's work that there was a geometry in which almost all of the Euclidean axioms hold, but the concept of distance has been weakened a little and it is no longer true that in a triangle with two equal angles the corresponding sides are equal. Again, Hilbert gave reasons why this was not an idle curiosity but a fact pregnant with implications for not only the geometry of surfaces but also the theory of numbers.

One might conjecture that Hilbert left the 5th of his Problems out of his talk only because it is technical, because it is certainly significant. The distinguished Norwegian mathematician Sophus Lie, an old friend of Klein's who had died in 1899, had spent much of his life studying geometric transformations, and had found that these came in standard types. This was the start of a long and complicated history, in which other mathematicians joined, of whom Wilhelm Killing in Germany and Élie Cartan in France are perhaps the best remembered.[1] To get Lie to write up his results Klein and Poincaré had sometimes recruited bright young mathematicians and sent them out to Lie, who worked in Leipzig and Christiania, to act as high quality scribes. The result was a series of books and papers in which Lie set out a new branch of mathematics with implications not only for geometry but also for the solution of partial differential equations. If we look ahead we can see that later generations of mathematicians were to re-write the theory, and the modern theory of Lie groups is fundamental to many branches of physics, notably particle physics.

The theory as it stood in 1900 bore the marks of its long and arduous creation. It was not clear what features of it were truly fundamental and which were accidental. It often happens in mathematics that a problem is solved by an unduly complicated method the first time it is solved, and later work produces more direct methods hidden, for what ever reason, from the first investigator. Hilbert focused on one specific feature of Lie's theory: his geometric transformations were required to be differentiable. Hilbert suspected that this was not necessary, and the transformations Lie

---

[1] See Hawkins (1982).

was interested in could be described in such a way that they had to be differentiable. It should be enough, he suggested, to assume that the transformations were continuous, and their differentiability would follow.

Hilbert selected topics from number theory for the printed text with the same sensitivity as in the lecture. Number 9 asked for a law generalising quadratic reciprocity to arbitrary powers. This was a mainstream problem on the agenda of any algebraic number theorist. Mathematicians, it is sometimes said, really are a people who count '1, 2, 3, many'. This is understandable. A problem may arise in a simple area, although it may not seem simple the first time it comes up. For example: 'I have thought of a number, and if three is added to twice the number, I have 11. What is the number?' The answer is, of course, that the number is 4. But the history of mathematics documents a long period before the answer would be found the way it is taught today in school: let the unknown number be x. Then $2x + 3 = 11$, so $2x = 8$, and $x = 4$. From the ancient Babylonian period (say 1700 BC) we have texts showing that people could solve what we would formulate as quadratic equations. Our school syllabus still teaches how this is done. But we do not go on to teach cubic equations, then those of degree 4, then those of degree 5, and so on. The mathematics profession does not contain people saying 'Ah, I'm a specialist in equations of degree 23, but you need a degree 24 person. I can refer you to one, but I can't help'. The reason is two-fold. As it happens, equations of degree 4 can be solved by algebraic formulae, but only special equations of degree 5 and above can be. But more to the point it would not be a useful theory if all you could do was slog through, degree by degree, learning different methods. At some point mathematicians look for a general theory. They cherish the low-dimensional cases, but they balance them with the general case. Each approach has its merits.

Hilbert's 10th Problem looks deceptively simple, and it may be that Hilbert himself was among the deceived. In its entirety it runs: 'Given a Diophantine equation with any number of unknown quantities and with rational integral numerical coefficients: To devise a process according to which it can be determined by a finite

Box 3.1 An example when continuity implies
differentiability

Consider the question: what functions f satisfy the equation
$f(x + y) = f(x) + f(y)$ and the condition $f(1) = 1$ for all x and
y? Surely the answer must be that only the function $f(x) = x$
itself has this property. It is reasonably easy to see that for integers m indeed $f(m) = m$ — start by noting that

$$f(2) = f(1 + 1) = f(1) + f(1) = 1 + 1 = 2.$$

The same argument shows that $f(mx) = mx$ Next observe that
$f(m/n) = m/n$ — start by noting that

$$f(1/2) + f(1/2) = f(1/2 + 1/2) = f(1) = 1$$

and so $f(1/2) = 1/2$. An argument known to every mathematician in 1900 would then conclude the argument by saying
that *if* the function f is continuous, then $f(m/n) = m/n$ for all
rational numbers m/n implies that $f(x) = x$ for all real numbers. So, if f is continuous it is the function $f(x) = x$, and that
function is most certainly differentiable.

Hamel used Zermelo's contentious Axiom of Choice in 1905 to
show that there are, however, discontinuous functions f which
satisfy the equation $f(x + y) = f(x) + f(y)$.

number of operations whether the equation is solvable in rational
integers.' A Diophantine equation is a polynomial equation with
(usually, and in this case) integer coefficients.[2] For example, the
equation $x^n + y^n = z^n$ for any fixed integer n in Fermat's Last
Theorem is a Diophantine equation. They are named after the
Greek mathematician Diophantus (third Century AD) who was
interested in problems of this kind.[3] So Hilbert is discussing this
whole class of equations (and in particular $x^n + y^n = z^n$ for any

---

[2] Recall that the concept of being a integer had been stretched by number theorists
over the nineteenth century until the familiar integers were only a special case.

[3] Naturally the real historical story is much more complicated, see Bashmakova
(1997).

fixed value of n). He asks for a process which will determine if a given equation has integer solutions. This process is presumably in some unspecified way to be uniform, although it might well begin by treating the equation in a way that depends on its degree. More importantly the process is to be finite.

This is another example of Hilbert's interest in finite decision processes. It will connect in due course with his interest in axiom schemes of various kinds, which he will also require to be finite, and difficulties with this concept will ultimately derail him. But in 1900 it probably seemed to him to be a good general problem: provide number theorists with a method of seeing if a specific equation that interests them has integer solutions at all. In view of his success with the basis problem in invariant theory, it is entirely possible that Hilbert would have been thinking of an abstract argument which might well say that a given problem did have integer solutions but was of no use at all for finding them. Or he might have had an analogy with Galois theory in mind. Such a process might be a solace to number theorists, but it would not put them out of business.

Another technical problem followed. Classical number theory from the eighteenth and nineteenth centuries had permitted mathematicians to know a lot about what integer can be written in the form, say, $2x^2 - 3xy + y^2$. (For example, if $x = 2$ and $y = 1$ then $2x^2 - 3xy + y^2 = 3$, so 3 can be written in this way.) Hilbert now said that the time had come to recognize this topic and give it an appropriate level of generality. The coefficients should be anything that counted as an integer. There should be any number of variables (not just 2). The solutions should either be the same sort of integers as the coefficients or fractions composed of such integers. This is a typical mathematician's problem. It has aspects of the '1, 2, 3, many' business mentioned earlier. It asks for some difficult work, the reward for which will be an overview of the field. And, as Hilbert knew very well, Minkowski had shown in a letter to Hurwitz (published in 1890) how quadratic forms can be classified over the rational numbers. He had given criteria which enable one to say when one form can be converted into another and back by means of a linear change of variables (analogous to changing coordinates in geometry). Minkowski's theory was, however, difficult and elaborate, and Hilbert both acknowledged his friend's work and called for it to improved upon.

With his 12th Problem Hilbert reached the middle of his talk, the apex of an arch. He became correspondingly loquacious, and it would seem that this was a Problem he cared about particularly. To be sure he had left the subject of number theory, so this was a Problem for other people, but it was a way to his heart. He called the Problem the extension of 'Kronecker's theorem on Abelian fields to any algebraic realm of rationality', and the slew of terms alerts us that this was a tour de force, one of those passages where the expert suddenly outpaces the audience and affords them not the pleasures of understanding but a glimpse of what it must be like to find easy and natural what many find completely beyond their grasp. Here Hilbert sketched a vision of how a branch of mathematics should go that no-one except perhaps his friend Minkowski could have immediately understood. Let us take it apart a little.

Our first problem with this Problem is that much of what Hilbert wrote is actually wrong.[4] No matter that, not for the first time, his history is somewhat over-simplified. The question of what to attribute to Kronecker by way of a proof is rather difficult, and the suggestion that the first valid proof was provided by Weber in 1886 is also false (Weber's mistake was only observed as late as 1979!). In fact, it seems that Hilbert himself was the first to prove the Kronecker–Weber theorem. The real problem is that what Hilbert conjectures is actually incorrect.

One starting point is the observation that algebraic numbers pose many fewer problems for our understanding than do transcendental ones. Algebraic numbers are the roots of polynomial equations with integer coefficients, and transcendental numbers are not, so algebraic numbers can be manipulated, even in a computer, as if they are honest numbers. Because algebraic numbers arising from a specific polynomial can be added together,

---

[4] See Schappacher (1998).

subtracted, multiplied and divided (except for 0) mathematicians say they form a field. This field clearly contains the field of all rational numbers, so it is said to be an extension of the field of all rational numbers.

Hilbert therefore suggested that Galois theory was the way to think of this sort of topic, because Galois theory applies every time you have two fields, one contained in the other. It produces for you a group, which carries information about how the fields are related, and so carries information about the number α and the polynomial equation that defined α. Groups are algebraic objects that capture and generalise the notion of symmetry (as we saw in Box 1.2) and the idea of combining transformations. They come in a variety of kinds, but ones where the multiplication is commutative ($x.y = y.x$) are easier to understand than ones where the multiplication is not commutative. Finite commutative groups are so well understood it is possible to write down a description of all of them. Hilbert used the synonym 'Abelian' for commutative, a term that derived rather obscurely from the interest of the Norwegian mathematician Niels Henrik Abel (who had died in 1829 at the age of 26) in the solution of polynomial equations.

The Kronecker–Weber theorem is the assertion that you get a commutative or Abelian extension of the rational numbers essentially if and only if you have adjoined roots of unity. This is surely a beautiful result. It connects a very simple class of group with a particular kind of extension, and it does so by invoking the values of, as it happens, a particularly well-known function. It says that from a certain perspective we know all about Abelian extensions of the rational numbers. Inspired by this result, Hilbert then took the simplest possible replacement of the field of rational numbers, the fields obtained by adjoining an imaginary square root, such as $\sqrt{-3}$. These fields are called the quadratic imaginary fields. He then asked about their Abelian extensions, and he suggested that they too come about by adjoining numbers that are the values of a certain function on the numbers of the quadratic imaginary field. In this case he had in mind the so-called elliptic modular function, usually denoted j.

This claim of Hilbert's is wrong. It is, so to speak, trivially wrong

Box 3.2 Algebraic numbers

Consider a solution of the polynomial equation $x^3 + x + 1 = 0$. It has only one real root, which we shall denote a, and which is a number between -1 and 0. Even if we know nothing else about this number we can do a lot of arithmetic with it. We can write down expressions like $2 - 3\alpha + 4\alpha^2$. We can discover that $1/\alpha = -\alpha^2 - 1$, and that $\alpha^5 = -\alpha^3 - \alpha^2 = 1 + \alpha - \alpha^2$. In fact, any polynomial expression in α with rational coefficients can be written uniquely in terms of rational numbers, α, and α^2. Indeed, any rational expression in α, by which we mean a quotient of one such polynomial by another, can likewise be written as a quotient of polynomials involving only the first and second powers of α.

Box 3.3 The Kronecker–Weber theorem

The Kronecker–Weber theorem is the assertion that a commutative Abelian extension of Q arises if and only if the adjoined numbers have the form $e^{\pi i x}$ where x is an element of Q, i.e. is a rational number. (A technical point is that perhaps more than one such number is to be adjoined).

For example, the algebraic number $e^{\pi i/5} = \xi$, which satisfies the equation $x^4 + x^3 + x^2 + x + 1 = 0$, gives rise to the field extension $Q(\xi)$, and that must be an Abelian extension. So the Abelian extensions, the simplest to understand, can be analysed by saying that they come about from the adjunction of a particularly simple class of algebraic numbers.

because, as the experts could see, you get Abelian extensions of a quadratic imaginary field by adjoining a root of unity, but such numbers are not always values of the *j*-function. But this was taken to be a lapse by the master, and so the conjecture that forms Problem 12 was taken to be that Abelian extensions of quadratic imaginary fields are obtained by adjoining either roots of unity (which is trivial) or values of the j-function at points of the quadratic imaginary field (which is interesting, surprising, profound, and to be proved). With hindsight this too is wrong, so the historian's question, which we shall discuss below, is to see whether the influence of Hilbert contributed to the confusion.

The 14th Problem was the only occasion Hilbert had to refer to invariant theory. He suggested that it might be interesting to restrict the group involved, and he then went on to propose a second problem which was to lie fallow for some time, and we shall refer to it later.

Problem 15 was one of those which could have cropped up in any account of the Problems of the day. A German mathematician called Schubert had taken up questions about configurations of curves. The oldest of these goes back to the Greek geometer Apollonius, and asks for the number of circles that touch three given circles. The answer is 8, according as the fourth circle encloses none, or only one, or exactly two, or all three of the given circles. Schubert produced answers to a wide variety of similar questions. For example, he considered quadric surfaces (surfaces defined by a quadratic equation in three variables, such as $x^2 - yz = 1$) , and came up with this magnificent number for the number of quadric surfaces touching 9 given quadric surfaces in three-dimensional space: 666 841 088. How could a mathematician resist such a number? But what gave many people a slight frisson of alarm was the methods by which Schubert claimed to find these numbers, which he called the method of conservation of number, and is illustrated by the example in Box 3.4.

Hilbert cautiously challenged his audience to establish these numbers rigorously, if possible by using Schubert's methods, and to characterize those situations where the methods work.

With Problem 17 Hilbert returned to topics he had worked on, and to another technical question that will find its proper place below. Problem 18 is in fact three rather different problems. All are loosely connected with the idea of how groups can act on spaces. The simplest and in some way the canonical example is the chess board, or rather, to be precise, the infinite chess board. We take the infinite board because, whenever we appreciate a specific pattern

Box 3.4 The Schubert calculus

How many lines are there that cross four given lines in space? Schubert argued that one could take a special case (see Figure), where the first and second lines cross (at a point *A*, say) and the third and fourth lines cross (at a point *B*, say). Notice that the two pairs of lines each define a plane. In this case the answer is that two lines meet the given four: the line α joining the points *A* and *B* and the line β common to two planes.

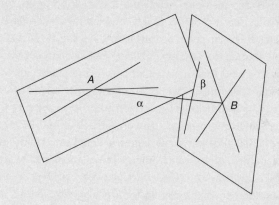

So, said Schubert, and this was the alarming step, because we found a finite answer here, and not an infinite one, the answer is that in general there will be the same finite number of lines meeting four lines given in space.

of this kind, on a floor or a wall, we usually forget about the edges; mathematicians do this by abolishing the boundary altogether. It is easy to imagine moving the entire board two squares horizontally, so that each white square lands on a white square, and each black square on a black square.

Although the board has moved, the pattern remains as it was. Such a move is called a symmetry of the board. There are many other symmetries. Apart from moving it any even number of squares horizontally we can also move it an even number of

Box 3.5 Mirror symmetry

In the late 1990s a number of physicists began to make conjectures based on their insights into how certain mathematical ideas work in physics, which had extraordinary implications for enumerative geometry. What they called mirror symmetry made precise estimates for such things as the number of rational curves of degree d on a 3-dimensional variety defined by an equation of degree 5 in 4-dimensional projective space (a smooth 3-dimensional variety defined by a homogeneous equation of degree 5 in 5 variables). Apparently it should have 2875 such lines (curves of degree 1), 609 250 conics (curves of degree 2), and, for example, 229 305 888 887 625 curves of degree 5. In 1998 the Russian mathematician Kontesevich was awarded a Fields medal for wide range of inter-related work, some of which contributed to a rigorous proof of some of the ideas of mirror symmetry, and in particular the number 242 467 530 000 of curves of degree 4. Schubert would have been delighted, and the rich interaction of mathematics and physics in this context would surely have delighted Hilbert too. [The numbers are taken from Cox and Katz (1999), p. 24]

Box 3.6 The infinite chess board

squares vertically. We can rotate the entire board through a right angle about the centre of a square. We can rotate the entire board through a half a turn about a corner of a square. We can compose the symmetries with each other and get more symmetries.

Mathematicians around 1900 were interested in such questions for a number of reasons. The collection of all symmetries of a pattern form a group, and group theory was taking off as an independent subject. It was known, for example, that in group theoretic terms there were precisely 17 different patterns that an infinite plane could have, and each group and its associated pattern were well understood. Classification theorems of this kind are useful in mathematics.

Poincaré had come up with a profound use of this idea, which connects the 18th Problem to the 22nd Problem. Consider the infinite chess board, but colour all the squares the same. Any movement of the board that places one square exactly on another defines a symmetry of the board. Now consider only what are called the translations, motions that move the entire board a number of squares horizontally and a (possibly different) number of squares vertically. Whatever translation is considered, points inside a square wind up inside a square. Points on an upper or lower edge wind up on an upper or lower edge; points on a right or left hand edge likewise wind up on a right or left hand edge. This suggests taking just one square and looking at what happens to its points. Interior ones remain on the inside, but it is impossible to say whether the points on an upper edge go to an upper or a lower edge. The problem of which edge to choose disappears if we glue the upper or lower edges together, and likewise the right and left hand edges. When this is done, the familiar figure of the torus (American doughnut or inner tube) appears. What Poincaré had done was to apply these ideas to the non-Euclidean plane, and to conjecture that almost every Riemann surface arises in this way. This is the astonishing uniformisation theorem that Hilbert made the subject of Problem 22.

The same question can be asked about the surface of a sphere. What tiles can it be given which form a pattern such that motions of the sphere are symmetries of the pattern? The answer is that the tiling may be that of a regular solid, or a truncated solid, or resemble a division of the faces into segments like an orange. Even in these cases the mathematics yields profound conclusions to which Felix Klein had devoted a whole book in 1884, and the regular solids continue to make significant appearances in mathematics right up to the present day.

Nor did mathematicians see any reason to stick to the Euclidean plane. For example, the generalisation to Euclidean 3-dimensional space is crystal lattices, which had been the subject of prolonged interest to chemists and also mathematicians throughout the nineteenth century.[5] Again it was found that only a finite number of cases can occur. Being mathematicians they naturally wondered about the general situation in any number of dimensions.

Hilbert directed attention to three questions in particular. First, he asked if it is the case that Euclidean n-dimensional space admits only finitely many patterns? Hilbert's second question was somewhat more technical. In all known examples at that time, a covering of the plane by tiles or of space by crystals is associated with a group. The group, as it were, moves the tiling around so as to realise the symmetries of the pattern. The tiles which form a given pattern are far from unique, and many ingenious designs, such as Escher's, come about by choosing them in a clever way. Hilbert now asked: are there any tiles which form a pattern covering the plane but which are not associated with a group?

Hilbert had also asked about what can be done when the regions do not fit together completely but leave gaps. How well can spheres of unit radius be packed together, for example, in the sense that the gaps are as small as possible? In the plane it is reasonably easy to show that the hexagonal arrangement is the best possible, although the first proof was given by Thue in 1910. Hilbert's passing remark that the problem is important in number theory was an allusion to Minkowski's work on the geometry of numbers, where the use of geometry enabled him to improve a technical result previously found by Hermite by algebraic means.

The run of Problems from 13 to 18 is the least coherent block.

Hilbert's touch was less sure here, and the influence of these Problems on the later development of mathematics has been less substantial, as we shall see.

However, with the final set of five Problems Hilbert was on surer ground. He had already discussed the celebrated Dirichlet problem in a talk in Munich the year before, and again in his lecture in Paris.[6] The printed paper set out a larger stall. How can it be, he asked in Problem 19, that a very general kind of partial differential equation has solutions that are the best behaved kind of function of several variables? This seemed to him so significant that it called for a general proof.

This question of investigating rigorously the mathematical results suggested by physics occupied the 23rd and last of his Problems too. A branch of mathematics called the calculus of variations investigates problems such as the curve of quickest descent with which Hilbert had illustrated the opening part of his lecture. Whenever possible, mathematicians had used the existing elementary calculus as a guide. The calculus teaches one how to find the maximum and minimum values of functions. The calculus of variations sets out to find the function which gives an integral its maximum and minimum values. But that question is much harder, and Hilbert knew very well that distinguished mathematicians had made quite elementary blunders. They had failed to give conditions under which the problem can definitely be solved. They had failed to give criteria whereby a maximum value can be distinguished from a minimum one. Weierstrass had made definite progress, and more recently a younger mathematician called Adolf Kneser was doing even better, but the situation was still unsatisfactory.

The problem was intimately connected to the Dirichlet problem, as every one in Hilbert's audience knew. Riemann had used an argument of this kind to show that a certain integral was minimised by a function in order to deduce properties of the function. He argued that the integral never took negative values, and it took values that could be arbitrarily small. Therefore—or so he said—it

---

[6] Although his published version falls well short of a proof. See Brezis and Browder (1998), p.85.

took the value zero for some function.[7] Weierstrass had objected. He took an entirely different integral, but he showed that the general argument was invalid. With his integral it was clear that it took arbitrarily small values, but clear that there was no function for which it took the value zero. It is rather like being asked to find the smallest positive number (zero is not allowed). Whatever number you pick, there is always a smaller one, and there is no smallest.

Hilbert wanted the calculus of variations to be re-written so that it was clear when one could argue as follows: to minimise the value of this integral, pick a sequence of functions that produce steadily smaller values of the integral. Arrange that these functions tend to a limiting function. Then that function minimises the value of the integral. In other words, find conditions in which a direct approach works and Weierstrass's objections are invalid.

The background: mathematics in 1900

Hilbert's paper was exceptional among the papers given at the Congress, as he had intended it to be. It is remarkable how many of these topics he had worked on himself or could argue were directly connected to his own work. As many as three-quarters of the ones he chose were related to what he had done or was about to do. Some, like the problems to do with the Riemann zeta function or the uniformisation theorem, had established themselves as major open problems it would have been odd to ignore. The more mainstream his choice was, the more impressive he looked. It is not entirely idle therefore to consider what the state of mathematics was at the time, what other problems he might have chosen, and what indeed some one else might have selected as the problems for the next century.

Mathematicians often regard their subject as dividing into three subjects: analysis, algebra, and geometry. Analysis includes investigations into the rigorous use of the calculus and the solution of differential equations, ordinary and partial (for functions of one or

---

[7]   Riemann's justification was somewhat better. He at least argued that the integral depended continuously on the function being integrated.

more variables). Algebra by 1900 was ceasing to look like high school algebra, and becoming more involved with the study of number-like objects, with linear transformations and matrices, and with structured objects like groups. Geometry could no longer be regarded as synonymous with the topic of Euclid's *Elements*, since non-Euclidean geometry had been accepted for forty years as a logically consistent, physically plausible alternative to Euclid. Instead it embraced a family of different geometries, of which projective geometry was arguably the dominant one.

Most of the other subjects in mathematics fitted around and within these three domains. Number theory in the style of Hilbert was closer to algebra, but number theory in the style of Minkowski was more analytic. Differential geometry, the study of curves and surfaces in space, belonged equally to analysis and geometry, while its cousin algebraic geometry gave the same themes an algebraic emphasis. Analysis itself had acquired an impressive theory of functions of a complex variable, within which the study of elliptic functions was still the central theme. Analysts were beginning to realise that the study of continuity, as opposed to differentiability, posed a vast and delicate collection of questions.

If there was a new concern that did not fit any of those fields, it was with the very foundations of mathematics. Immanuel Kant had grounded mathematics in intuitions which enabled us to deal with the discrete and the continuous. Discrete quantities can be counted, continuous quantities can be measured. The problem with this tidy, and traditional, approach was that it proved incapable of yielding the precision that the rigorous calculus required. What worked was what we have already seen called the arithmetisation of analysis. This reduced the calculus to the study of limiting processes, and explained real numbers in terms of families or sets of rational numbers. Since rational numbers are easy to explain in terms of integers, this left the calculus as the study of limiting processes (its characteristic property), integers, and sets of integers. This was, for example, how Dedekind and Cantor regarded it in their different ways. But more recently there had been some disquieting developments. It seemed to some that it was not necessary to retain the integers, and that they too could be explained as

Box 3.7 Projective geometry

Projective geometry is the study of the properties figures have in common with their shadows. So three points lying on a line is a projective property, and three lines meeting in a point is another, but distance between points, and the size of angles between intersecting lines, are not projective properties. Any projective property is a property of Euclidean geometry, so projective geometry is more basic than Euclidean geometry.

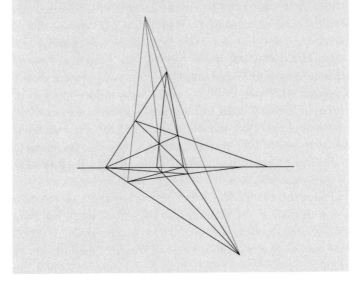

certain kinds of set. On the other hand, as we shall see later, problems were emerging with the original naïve conception of a set.

All these problems were fuelled by the anxiety occasioned by the discovery of non-Euclidean geometry. How could it be, mathematicians asked themselves, that they had been so wrong for so long? Two thousand years of suggesting that geometry was a logically compelling description of the world, the very paradigm of a logical description of something worth having, had to collapse when another geometry was found with an equally good logical structure but different theorems. Once the logical structure of Euclid's *Elements* was shown to have flaws, it was imperative to provide mathematics with a genuinely rigorous collection of arguments.

It could not be enough, for example, to say simply that mathematics works, that physicists use it and it doesn't let them down. Not only would this hand the reigns over to the physicists, who by 1900 were a distinct collection of people in other departments with other concerns, and mathematicians were not about to relinquish their privileged positions in the intellectual hierarchy. It would also be hugely untidy, for the plain fact was that mathematics and physics did not fit together well. Physicists often wanted results that mathematicians could not provide (the Dirichlet principle was a case in point). But worse, when the mathematicians sorted the problem out to their satisfaction the theorems were often not what the physicists wanted to hear. Physicists believed that something called the Dirichlet principle guaranteed the solution of the Dirichlet problem. Mathematicians showed that the Dirichlet principle was unsound, and that even when the Dirichlet problem could be solved its solutions might have quite different properties from what the physicists expected. The major theorists of light, electricity and magnetism (Hertz and Lorentz among them) had produced theories which Poincaré had just shown to be inconsistent not only with each other but, in Lorentz's case, with such basic beliefs of the physicists as Newton's laws of motion.

Hilbert, it is clear, had touched almost all the bases. He had directly addressed issues in the foundations of mathematics, raising problems concerning Cantor's theory of sets, and the nature of the concept of distance. He stressed the need for rigorous general theories. He had addressed head-on the imperfections in the relationship between mathematics and physics. He chose problems from analysis, algebra, and geometry. It is only within the individual domains of mathematics that he missed topics, and that was inevitable and wise. A list of 123 Problems would surely have been worthless, better to claim significance for some and make no attempt at completeness. Better too to speak from strength and raise problems which were close to his own work.

Roads not taken

That said, there are some interesting omissions, which shed light on the sort of mathematician Hilbert was. Hilbert might have mentioned the problem of defining what an integral of a function is. This is a problem that goes back to the invention of the calculus, and had been re-opened by Riemann and Cantor in the 1850s and 1860s. It is urgent when the function to be integrated is not even continuous. Such functions can cause problems when they are represented as sums of trigonometric functions, a process called harmonic analysis. This is a hugely useful process, but it does not always work. It is possible to write down the best possible trigonometric sum for a given function and discover that the sum and the function do not take the same values at the same points. The process of obtaining this trigonometric sum relies heavily on integration, so the best way of defining the integral was a topic of lively investigation, but one that Hilbert chose not to mention.

We might also observe that while he confidently calls for a theory of quadratic forms in any number of variables, he is almost silent on the need to develop a theory of functions of several variables. In particular, there is a very close tie between functions of a single complex variable and harmonic functions of two real variables, but this connection breaks down when the number of variables increases. In fact, in 1900 there was not much by way of a theory of harmonic analysis for functions of several variables at all, another topic that Hilbert slid over. There was even less of a theory of complex functions of several variables, but again Hilbert was silent. He also shared the widespread feeling that differential geometry, but not algebraic geometry, was about curves and surfaces, but not higher-dimensional objects.

And while Hilbert was pleased to invoke the name of Riemann on many occasions, he did not single out one of his most innovative and fecund ideas, which by 1900 was beginning to make its way under the name of topology. Riemann had shown that the theory of complex functions depended on the nature of the domain of the variable: was it a sphere, or an algebraic curve of some kind? In particular he showed that the most important fact about the domain of a complex function was given by a non-negative number called the genus (and which is illustrated in Box 3.8). Topological ideas were spreading to other fields, and Hilbert was to encounter topological problems in his own work, but it remains a theme that he may be said to have underestimated. It is present in some of his Problems, but it is not the organising principle it was later to become for much of mathematics.

Many in the audience and the larger mathematical community must have begun to make comparisons between Hilbert and Poincaré. As long as the exercise is not taken too seriously, it has a point. If Poincaré had thought to do something very like what Hilbert did, just by drawing on topics he had worked on he would have been able to bring up the uniformisation theorem, as Hilbert did, but he could then have made a stronger case for group theory, which he regarded as a truly fundamental discipline. Then Poincaré might have raised the need for a theory of functions of several complex variables. He had already established some fundamental properties, and was to go on to find striking ways in which the several variable theory differs from the theory of a single variable. He could have directed attention to the geometric theory of real differential equations and its implications for celestial mechanics. In that subject he had exposed one of the first major examples of chaotic behavior and the sensitive dependence on initial conditions on the future evolution of a dynamical system. Hilbert did mention this, but it has the air of politeness about it whereas Poincaré could have spoken with real assurance. Poincaré could then have gone on to raise the need for a topological analysis of manifolds of dimensions greater than two. He had been led to this topic because his work on celestial mechanics involved several variables, and in this matter he was more Riemannian than Hilbert. Poincaré also had strong views on the relationship between mathematics and physics, which he had studied in depth and written about at length.

The curious thing is that we can make a detailed comparison between Poincaré and Hilbert as expositors at International Congresses of Mathematicians, and when we do so Poincaré does not say these trenchant things. He is much more gentle, and the

Box 3.8 The genus

A wide class of surfaces is classified by their genus, which can be defined mathematically quite easiy but is much easier to illustrate. The familiar sphere has genus 0. The usual (or one-holed torus) has genus 1, and so on as the figure illustrates.

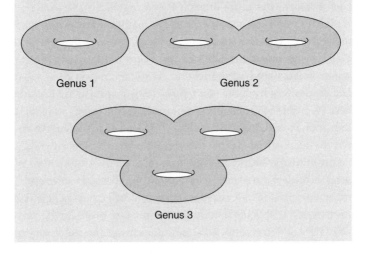

Genus 1   Genus 2

Genus 3

comparison shows us something else about how well Hilbert went about his task.

Poincaré's address of 1897

Poincaré gave a general address to the first ICM in Zurich, which Minkowski was to find bland and unexceptional.[8] He began it by asking, rhetorically, what is the use of mathematics? Dismissing with a confidence that would be out of place today the mundane desire to make money, he asserted that mathematics had a triple purpose: to aid in the understanding of nature, to help the philosopher make precise notions of number, space and time; and also an aesthetic purpose. Indeed, this aesthetic element is, he asserted, the very means by which mathematics and physics advance inseparably together.

---

[8] Poincaré (1898).

It is no longer possible to believe, he said, that pure reflection will yield a law of nature; all attempts in that direction have failed. Laws of nature are drawn from experiment, but they are expressed in a special language, which is created by the analyst. Moreover, unlike the results of this or that experiment, which are particular, a law is general. Experiments are approximate, a law is exact, or at least pretends to be. So to formulate a law one has to generalise; but how? Every truth can be generalised in infinitely many ways. The only possibility is by means of analogy. So, for example, in passing from Kepler's laws to Newton's, both theories agree that a single planet travels in an ellipse. However, in the new theory it becomes possible to study the orbits of the planets as they perturb each other, even though no-one has written down an equation for such a path, whereas had one remained with Kepler's laws it would have been necessary to treat the perturbed orbits as some sort of gener-alisation of an ellipse. Similarly, Maxwell's equations were written down, said Poincaré, because their author wished to present a symmetrical theory, 20 years before there was an experimental justification. Or again, the vast domain of potential theory embraces the electrostatic potential, magnetism, and the theory of heat. The analogy this suggests provokes scientific discovery, marked by the electricians adopting the terms flux of force from the hydrodynamicists.

On the other hand, analysis owes much to the study of nature. It not only raises problems and forces us to choose between them, it raises problems we would never otherwise think of. Nature is much more imaginative than Man. But it does more, it aids us in solving the problems, sometimes by suggesting the solution, some-times by suggesting the reasoning by which the solution can be found. In the first case, physics makes clear that a series represent-ing the solution formally must converge, even when a rigorous proof is lacking. In the second case, the behaviour of a quantity represented by a harmonic function may literally show that a cer-tain analytic function exists. Poincaré referred to Klein's little book on algebraic functions and their integrals[9] but he could equally

---

9   Klein (1882).

well have referred to his own (more demanding) paper on the partial differential equations of mathematical physics, where he had allowed that it was usual for physicists to rely on more or less intuitive proofs of the existence of solutions to these differential equations, based on an appeal to Dirichlet's principle.[10] Such observations, he said, although without value for the mathematician, are of the right sort to satisfy a physicist because they leave the mechanism of the phenomena apparent. However, the more rigorous mathematical arguments for the existence of solutions, such as his own, depended on convergence arguments. This convergence was usually too slow, and the approximations involved too complicated for such approaches to yield effective numerical procedures. I think one might argue here that Poincaré at the very least would have welcomed a transcription of the 'mechanism of the phenomena' into the mathematical proof of Dirichlet's principle, in keeping with his juxtaposition of the roles of laws of nature and theorems.

These examples left Poincaré with a seeming paradox: could there be two kinds of rigour? In one of his more technical papers[11] he had argued that one could not be content with the lack of a rigorous proof. Analysis itself should be able to solve such problems, and any rigorous solution is, of course, a solution; even if crude it nonetheless teaches us something. It was not needless pedantry to seek the rigorous solution of equations established only by approximate methods and which rested on imprecise experimental foundations, because how else could one be sure that something less than a rigorous proof was not actually flawed? Nor had one the right to say that something inadequate for mathematics was yet good enough for physics—the line was impossible to draw. One could not, as a mathematician, settle for less, and in any case many of these equations had applications not only in physics but also in pure mathematics (for example, Riemann himself had based his magnificent theory of Abelian functions on his use of Dirichlet's principle).

---

[10] Poincaré (1890).

[11] Poincaré (1890).

But if the canon of rigour was something to strive for, the fact remained that physicists successfully settled for less. In his ICM address Poincaré resolved the paradox by asserting that almost always the problem in providing a rigorous mathematical proof came down to establishing the existence of certain limits. But physicists deal with quantities known only approximately, so they may always pass to a better behaved function, amenable to rigorous mathematics arguments. In this way the paradox disappears. This turned out to be a fallacious argument, as is shown in Box 3.9.

Poincaré and electromagnetic theory

It is worth while amplifying these reflections of Poincaré by considering some aspects of his own work in electromagnetic theory. Throughout the 1890s and the first decade of this century, Poincaré pursued interests in mathematical physics. The first edition of *Électricité et Optique*, which came out in 1890 and dealt with the ideas of Maxwell, Helmholtz, and Hertz, was followed in 1894 by a further account of Hertz's ideas. In 1906, Poincaré introduced what are called Poincaré stresses; non-electrical forces designed to hold the electron together, and which enable the conservation of energy and momentum laws to be applied consistently in the determination of the electromagnetic mass of the electron.

Poincaré, however, was only willing to introduce physical hypotheses when the mathematics required. Otherwise, in his *Électricité et Optique*, he argued that a physical system would admit infinitely many mathematical descriptions that agree with all experiments, and in so doing he felt he was true to Maxwell's original ideas. At this stage he was neutral on the question of the existence of atoms, for example. He made mathematical sense of the approaches of Maxwell, Helmholtz, and Hertz by regarding certain conservation laws as fundamental, and emphasised Newton's laws, the relativity of motion, and the principle of least action. The result was a clean, logical textbook aimed at uniting experimental facts and mathematical principles, but it did not succeed. In the late 1890s, his analysis revealed a contradiction between experimental facts in optics and Newton's third law. Poincaré's preference was

---

Box 3.9   Hadamard's counter-example

In 1917, Hadamard gave a lecture at Zürich, where he refuted the view of most mathematicians that because any function can be well approximated by analytic functions, a non-analytic partial differential equation could be replaced by a 'nearby' analytic one, with only slight effect on the solutions. As he put it: 'in my opinion this objection would not apply, the question not being whether such an approximation would alter the data very little, but whether it would alter the solution very little.' Then followed his example.

The partial differential equation

$$\frac{\partial^2 u}{\partial x^2} + \frac{\partial^2 u}{\partial y^2} = 0$$

with the initial data $u(0, y) = 0$ and

$$\frac{\partial u}{\partial x}(0, y) = u_1(y) = A_n \sin(ny)$$

where A_n becomes very small as n grows large (for example, $A_n = 1/n^p$). The data can be made as small as you wish. But the equation has as its solution $u = \frac{A_n}{n} \sin(ny) \sinh(nx)$, which is very large for any value of x other than $x = 0$.

---

markedly for Newton's law, but he had no resolution of the contradiction.

Poincaré's address of 1908

In 1908, in his address to the International Congress of Mathematicians in Rome on the future of mathematics, Poincaré returned to these themes of his address of 1897.[12] He rejected the idea that mathematics should be done solely at the behest of physicists, on the grounds that were that to be the case, mathematicians

---

[12] Poincaré (1908)

would stand unarmed before scientists. But they, in their turn, have not waited and do not wait upon the call of utility, but search for laws. Laws group together otherwise isolated facts by analogy, facts which, however striking, would be sterile were it not for the eye of the expert who discerns which facts conceal interesting general ideas. Likewise the mathematician structures the millions of mathematical facts around mathematical theorems.

The important laws in either case order the facts around them; the analogy permits one to decide which facts are alike, which differ significantly. The most important facts belong to the very general laws of physics. The mathematicians who praise a theorem or a proof for its elegance likewise prefer theorems and proofs that permit them to view a whole field at a glance, which enable them to say *why* something is the case (as a lengthy calculation does not). This economy of thought, which, Poincaré reminds us, Ernst Mach had regarded as the role of science, depends on the clarity afforded by the discovery of general laws.

Laws in physics, and theorems in mathematics, are often a place where new words can be found, which are the names of the organising principle. Energy, as in the conservation of energy, is one such word. For mathematics, Poincaré proposed his favourite topic, that of the group, in particular, the idea of isomorphic groups, which dispensed with matter and retained only form.

Mathematics, according to Poincaré, must always move in two opposing directions: that of critical self-reflection and towards the study of nature. The first leads, rightly, to such matters as the study of postulates, of unusual geometries, and of functions with strange properties (all these Poincaré knew well). The second invites too naïve an answer, which may not even be available (such as an explicit formula for a solution to a differential equation). Better, and perhaps easier to find, is the qualitative answer, after which comes a quantitative one, perhaps to some level of approximation. Here Poincaré gave vent to what became a famous aphorism: 'There are not solved and unsolved problems, there are only problems which are more or less solved.' Like much of Shakespeare, this has been quoted out of context (which does not make it wrong). The context here is that of the power series solution, which may converge too slowly, or whose very convergence may be the key to the puzzle, and where successive terms may or may not be generated by a simple law. Again, Poincaré had had intimate acquaintance with all these possibilities.

Poincaré then turned to specifics. In arithmetic he was struck by the analogies between the theory of congruences and that of algebraic equations. This refers, obscurely, to what we now call the theory of ideals, which was the topic he selected in algebra itself—the analogy between ordinary integers and algebraic integers. In geometry he picked out the analogy that enabled one to speak geometrically at all about spaces of dimensions greater than 3. At this point he returned to the qualitative–quantitative distinction, singling out Riemann's introduction of topology. Then we get Cantorian set theory, where, as I have shown elsewhere, Poincaré did *not* say that set theory is a disease, and finally we get a nod towards Hilbert's study of postulates.[13] Poincaré praised the novelty in Hilbert's work, his geometries that were not based on the idea of a number manifold, and for which Lie's transformation groups could play only a secondary role. For once, even Poincaré could not offer a psychological account of geometry, only lament its absence.

One might well feel that Poincaré's offering in 1908 is something of a soufflé. Above all, it lacks precise suggestions as to what you might do if you were caught by the message. Poincaré's address, although thoughtful and drawing on his own fund of experience, is vague. It offers nothing that one can do in its spirit, nothing that would be a contribution to mathematics à la Poincaré. Doubtless it was not meant to lead in that way. Hilbert's talk, in contrast, offered anyone who read it the opportunity to do something as a direct response.

The comparison with this talk by Poincaré sheds one fresh shaft of light on Hilbert's problems. When discussing Cantorian set theory, Poincaré remarked upon certain paradoxes and apparent contradictions that would have delighted Zeno of Elea and the school of Megara. Each of us, he said, must look for the remedy, but whatever the remedy adopted, we can feel the joy of a doctor called to a

---

[13] This is the starting point for the entirely erroneous but often repeated claim that Poincaré said set theory is a disease, see Gray (1991).

Box 3.10 Two paradoxes

The Burali–Forti paradox. On the one hand, given any ordinal it is possible to define a larger one. Therefore there is no largest ordinal number. On the other hand the set of all ordinals is itself an ordinal, and must be larger that any ordinal. The paradox can be struck down by observing that not everything that looks like a set actually is a set. The challenge is then to keep mathematics based on something like sets, which are intuitively simple and basic, capable of yielding precise definitions of numbers and other fundamental concepts, while avoiding further paradoxes.

The Richard paradox arises by asking: what is the smallest positive integer that cannot be defined in less than a hundred words in English? The number of sentences in English of 100 words or less is finite (if huge) and so only finitely many positive integers can be defined in this way. That leaves infinitely many, of which one must be the smallest, so there is such a number. But then it is defined as 'the smallest positive integer that cannot be defined in less than a hundred words in English', and that is plainly a sentence of less than 100 words. This paradox raises questions about what sentences may be said to define anything. Its resolution may be found in an analysis of language, rather than mathematics.

beautiful pathological case. These few words do not amount to a rejection of set theory, the Paradise from which Hilbert said mathematicians would never be driven. They point to a problem which Poincaré is plainly optimistic can be solved. His own remedy was to avoid introducing objects that cannot be defined in a finite number of words. But the existence of the problem, and its growing visibility, shows that, perhaps inadvertently, Hilbert had picked for his first problem a topic with a lot of life in it.

The paper Hilbert presented in 1900 is not merely a string of good, difficult problems. It contains a series of arguments, strung like beads on a necklace, about why you should care, and why you will have done significant work if you make real progress on any of them. They will lead to theories that shed light on the topic, and on the topics around them. They will generate more problems. They will enforce high standards of rigour, necessary for the proper comprehension of mathematics itself. Finally, they are a challenge, which simply has to be met, and which ultimately one must be confident of meeting.

CHAPTER 4

The early response: 1900–1914/18

Hilbert and the viability of the Hilbert Problems

The Hilbert Problems were to be a great success, more than Hilbert could reasonably have predicted. Once brought to life by Hilbert they then lived a life of their own: growing up under his tutelage in Göttingen, moving abroad, and, in his old age, turning out unexpectedly. The children of his extraordinary optimism, some grew as he had predicted, while some were to turn obstinate, even recalcitrant. The insights on which they rested were not always sound, as Hilbert was painfully to discover. This outcome was by no means certain, and it requires some careful explanation.

Hilbert himself had gone to great lengths to present the Problems well. Other factors also helped to promote them. Felix Klein, with the support of Friedrich Althoff in the Ministry of Education, had built up a first-rate Mathematical Institute, which was by 1900 the leading place for mathematics in the world. It had a large and strong permanent staff, and a stream of visitors from many different countries. What Göttingen approved, the world of mathematics would listen to. Hilbert himself played an vital role in this. He was accessible, by the standards of German professors. The regular Wednesday walks in the hills were an opportunity to talk with him—usually about his own work—and he would intervene in the weekly seminars. He was, oddly enough, not a mathematician who grasped other mathematician's ideas quickly, and sometimes the seminar would grind to a halt as one or another person would try to persuade the master of the truth that was clear to every one else. Sometimes they would succeed, and sometimes it would become apparent that he had spotted a mistake . . .

He also supervised a remarkable number of students, 69 in all. German Universities had for some time realised the need to train graduate students in research. They set aside a room for journals, actively purchased the relevant books, organised research seminars for staff and students to work through recent significant discoveries. Professors expected to provide their students with problems for their theses, always hoping of course that this would not be necessary. The French at this time, for example, were much less organised. Hadamard's opinion, for example, was that while he was ready to supply a student with a problem to work on, he preferred the good students to come up with problems of their own.[1] Poincaré never managed to acquire students or followers in any number.

What Hilbert managed was to secure a virtuous circle in which his general ideas were taken up by first-rate students and refined, thereby contributing to the further advance of the ideas. The pioneer in this respect was, again, Klein. After the breakdown of his health he had shifted over to collaborative work, and by the time he was in charge of the Institute at Göttingen he was empire-building full-time. In the course of charming local industrialists to support his projects for genuinely applied mathematics, intervening in debates about the nature of mathematics education, organising what became a 23–volume *Encyclopaedia of Mathematics*[2], and being editor-in-chief of *Mathematische Annalen*, he also wrote survey articles, travelled extensively, and even did some research of his own. But his preferred way of working was to talk to his former student Robert Fricke, who from 1890 was a Professor in Braunschweig. In that way two different two-volume works were produced, written by Fricke, but a joint effort nonetheless. The topic was complex function theory, to be precise the theory of

---

[1] Quoted in Maz'ya and Shaposhnikova (1998), p. 190 from Hadamard 'Le rôle de l'inconscient dans la recherche scientifique', *Atomes* 26, 168.

[2] *Encyklopadie der Mathematischen Wissenschaften mit Einschluss ihrer Anwendungen.* (*Encyclopedia of Mathematical Sciences including their Applications*).

modular functions and the theory of automorphic functions, the topics that had brought Klein into contact with Poincaré, and which for him exemplified the underlying unity of mathematics.

Hilbert's way with his students was different, not least because Klein had always been a little weak on points of detail. Klein liked to think about the 'big picture' and to let others sweat out the awkward special cases. It would be some years before Hilbert would let up in that way. He did not, one supposes, do the student's job for them, but he had high standards for himself, although, as we shall see, this did not prevent him from falling into a variety of errors. His students, however, had to cope with the fact that their chosen area of mathematics might be one that Hilbert had 'left'. For all his involvement in number theory, he had no students in that subject until he had abandoned it himself. The first of these was Rudolf Fueter, who presented his thesis in 1903.[3] Then came Erich Hecke, in 1910, who went on to become a leading number theorist and a dominant figure in German mathematics after the Second World War.

Around 1900 he had a number of students working on problems in the foundations of geometry, of whom Max Dehn was to become the most distinguished. Another was an American called Anne Lucy Bosworth, who wrote on how Hilbert's segment arithmetic could be established independently of the parallel axiom— in fact Hilbert was in the forefront of the struggle to admit women to University. Another student in geometry was Georg Hamel, who wrote on Hilbert's fourth Problem. Matters were easier when Hilbert moved on from geometry to his next field, a branch of analysis that it is nowadays known as Hilbert space theory. This was to become the branch of mathematics underlying quantum mechanics, and if it owes much of its inspiration to Hilbert himself it also owes a great deal to some of his students: Erhard Schmidt (who made it more intelligible by systematically introducing geometric ideas), Ernst Hellinger, Hermann Weyl, and Alfred Haar. The difference between one professor, however good, and a professor and four students in regular contact, at least one of whom

---

[3] Fueter went on to become the first President of the Swiss Mathematical Society.

(Weyl) was of Hilbert's calibre, is obvious. Nowhere else in the world was there such a coordinated focus of research, and Hilbert space was a great success.

Hilbert did not push the Hilbert Problems on his students. Nor did he work extensively on the Problems themselves. As we shall see, he did pursue some of them, but not obsessively. Some of his students took them up, others did not, and in any case they were good problems. Hilbert space, for example, was attractive to bright ambitious students because it had the master's full attention, but it did not appear among the Hilbert problems. The Dirichlet problem did, of course, and so did the calculus of variations, and half a dozen students wrote in those areas, including a number of American mathematicians who took the topics back home with them. Others came to Göttingen although registered elsewhere, and this was another of the ways the Problems prospered: they offered another way in which ambitious mathematicians could associate themselves with Göttingen and benefit from its high standing.

In the end the Problems have the history they do for the usual mixture of reasons: their intrinsic qualities and people's opinions about them. As ever, the truth, lies, with the Devil, in the details.

There are fashions in mathematics, shaped partly by circumstance and partly by leading individuals. These fashions may be more obvious in the high-rolling world of mathematics education, with the notorious example of the 'New Maths', but they reach right up to the highest levels of research. Hilbert was not reflecting an orthodoxy when he gave number theory such a prominent place in Paris, he was presenting a German point of view to a French audience. He was not delivering a commonplace when he talked about the relationship between mathematics and physics, he was offering a programme of work to a community that was separating itself off from the physics community.

The nature of the connection between mathematics and physics had become fraught during the nineteenth century. By 1900 there were very few, if any, who commanded equal respect in both fields; Poincaré might have been the only candidate.[4] The nature of

---

4   Over the years, Mittag-Leffler attempted many times to persuade the Nobel Prize Committee to award the Prize for physics to Poincaré. His relationships

Hilbert's contribution was to call for special comment as the years went by because it was not, in the eyes of physicists, mathematical physics. His friend Minkowski had spent some time in physics laboratories, but he was exceptional (and at least to Hilbert he jokingly apologised for it, writing in December 1890 from Bonn that 'I have devoted myself for the time being completely to magic, that is to say, to physics. . . . Starting next weekend, I'll work some days every week in a blue smock in an institute that produces physical instruments; this is a kind of practical training and you could not even imagine a more shameful one').[5] In his obituary of Minkowski, Hilbert said that during his years in Bonn Minkowski felt closer to Hertz and his work than anything else, and had it not been for Hertz's early death would have devoted himself exclusively to physics.[6]

Hilbert, like Klein, felt it as his duty to lead from the front. They were a formidable combination, one in complete command of the administrative side, the other of the mathematics. Other mathematicians felt differently. The triumvirate of Kronecker, Kummer, and Weierstrass in Berlin had built up a large school, figures of 200 or more are spoken about for some advanced lecture courses. They had produced numerous doctoral students and placed them in Universities across Germany. But their successors lacked the institutional savvy of their forbears, and the place sank back a little. Among the French, Charles Hermite 'as distinguished a mathematician as he was, did not have the independence of a natural leader needed to create and maintain his own school'—in the

with them were poor, however, and he never succeeded. Between 1904 and 1909, Poincaré was nominated several times for the Nobel Prize in Physics on Darboux's initiative. In 1910, Mittag-Leffler organised another attempt, advising his nominators (Appell, Darboux and Fredholm) to avoid 'mathematics and refer to pure theory' because 'like those who are only experimentalists, members of the Nobel committee for Physics are crazily frightened by Mathematics.' Then Mittag-Leffler sent a circular round the entire physics community asking them to nominate Poincaré for the Nobel Prize, and thirty-four of them did. Rutherford was among those who did not, because he thought Poincaré was not enough of a physicist. The prize, however, went to the experimentalist Van der Waals. See Nabonnand (1999).

[5] Quoted in Hilbert's long and moving obituary of Minkowski.

[6] Hilbert (1909).

opinion of Klein.[7] He did, however, exercise a considerable influence behind the scenes in the hothouse Parisian mathematical world. Poincaré was happy to let his work speak for itself, but there was a profusion of that: his Collected Papers ran to 11 volumes, and there were many books, including four volumes of popular essays. Other national communities grew in the twentieth century, but it takes time (and fate) to rival the leaders. The Italian community was particularly strong, and perhaps the most independent. The British community still largely meant Cambridge, and in 1900 was in a quiet period after the death of Cayley and before the arrival of Hardy and Littlewood. The American community was growing fast, but it was heavily influenced by its connections to Germany (chiefly, but not exclusively Göttingen).[8] Hilbert and Klein were without peers when it came to a wish to dominate the mathematics community from a strong institutional base.

As we have seen, Hilbert presented Problems in four areas of mathematics. The first was broadly concerned with the foundations of mathematics; the last analysis and especially with the way pure mathematics can contribute to applied mathematics. These two proved to be major themes in the subsequent history of the Problems, and it seems best to take them in turn. The second of Hilbert's themes was number theory. Although this can easily become technical, it is also the topic where Hilbert's touch was most sure, and which provides us with the clearest account of his influence and the status that attached to the Problems. I shall therefore take it next. Last come problems that are loosely algebraic or geometric, and which have flourished less well. They will treated as a counter-point to the number-theoretic theme.[9]

---

[7] Klein (1926), 1, p. 292.

[8] The American–German connection is described in Parshall and Rowe (1994).

[9] I wish at this point to draw particular attention to the two anthologies devoted to the Hilbert Problems: Alexandrov, P. S. ed. (1979) and Browder, F. ed. (1976). A number of distinguished mathematicians wrote essays from a variety of points of view, and it would have been impossible to write this book without them. May this blanket acknowledgement replace what would otherwise be a festooning of the text with footnotes. Despite their age, these two surveys will remain the first port of call for the expert seeking to know how the Hilbert Problems stand for many years to come.

Hilbert's Problems on the foundations of mathematics

Hilbert's 3rd Problem

Hilbert's 3rd Problem was solved even before the address was published, as Hilbert commented in a footnote. This was one of his student Max Dehn's earliest achievements, and marked the start of a highly successful career in mathematics. Dehn went on to do work of central importance in the interaction between topology and group theory, and in his study of the origins of topology Moritz Epple has characterised him as an exemplary figure in the growth of abstract pure mathematics in the twentieth century.[10]

The 3rd Problem is by no means the most profound question in the list, but it illustrates nicely the sorts of concerns Hilbert's programme for geometry was throwing up. Consider first plane Euclidean geometry. In this geometry, the area of a triangle of a given height is proportional to its base (from which we deduce the formula area = one half base times height). The usual proof is given by Euclid in his *Elements*, and a sketch goes along these lines. Make a congruent copy of the triangle, and arrange it so that the two triangles form a parallelogram, as in Figure 1. Cut off a right-angled triangle on one side and attach it at the other, as in Figure 2, so as to form a rectangle. The area of a rectangle is indeed proportional to its base (and, as we would say, is given by the formula area = base times height).

It is reasonably easy to see what you would have to do to turn such a sketch into a rigorous cut-and-paste argument. One needs to be able to talk about congruent triangles and to construct them in any position. Properties of parallel lines are needed—and it was well-known by 1900 that in non-Euclidean geometry, where the properties of parallel lines are very different, the area of a triangle is not given by the usual formula. Of course, when one gets down to the details maybe other matters will present themselves, but it seems clear that in anything like plane Euclidean geometry there

---

[10] Epple (1999).

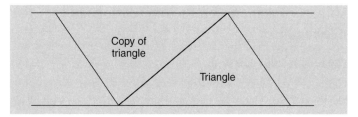

Figure 1

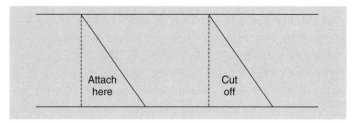

Figure 2

should be a theory of area that does not involve the calculus. This might then extend to geometries based on novel number systems, such as the non-Archimedean ones Hilbert had presented in his *Grundlagen der Geometrie.*

What Hilbert reminded his audience was that in *three-* dimensional Euclidean geometry there was not even a sketch of a calculus-free argument to show that the formula for the volume of a tetrahedron is of a given height is proportional to its base (whence the formula: volume = one-third base times height). All proofs involved the calculus, none proceeded by a cut-and-paste argument. This does mean there was no such proof, but the task had been contemplated unsuccessfully by a number of mathematicians and a positive outcome was unlikely. Hilbert conjectured that in fact the problem was impossible. The question then becomes to arrange for two solids of equal volume neither of which can be cut up into finitely many pieces and re-arranged so as to form the other.

Once the calculus is required, the theory of volume on Archimedean and non-Archimedean geometries could be very different, and that is where Dehn came in, for his doctoral thesis, written under Hilbert's supervision and published in 1902, was precisely on non-Archimedean geometry. What Dehn showed in 1902 was that indeed a tetrahedron and a cube of equal volume cannot be cut up and pasted together differently so that each turns into the other. The key observation in Dehn's proof is that, in addition to volume, there is a second numerical quantity associated with a solid that a cut-and-paste operation does not change. This, essentially, is obtained from the list of the edges and the dihedral angles (the angles along the edges between adjacent faces of the solid). What Dehn exploited is the observation that a cut and stick operation cannot change both the length of an edge and the sum of the dihedral angles at that edge. Some cuts may change the lengths, some the angles. Dehn found an ingenious combination which never changed. (One multiplies each edge length by the dihedral angle, multiplies this by a number one chooses carefully, and adds these numbers together.) Such a number is today called a Dehn invariant of the polyhedron.

Now, the dihedral angles of the regular tetrahedron are all $\phi = \arccos(1/3)$ (as can be seen by calculating the angles in triangle CED in Figure 3) while those of the cube are plainly $\pi/2$. It is possible to define a Dehn invariant which takes the value zero for the cube but a non-zero value for the tetrahedron. The result is that no sequence of cut and stick operations can cut a cube into pieces and reassemble it as a regular tetrahedron.

Dehn's proof was greatly simplified by Kagan the next year, when he set about describing Dehn's solution for a popular Russian mathematical journal.[11] But in fact the problem had almost been solved four years before it was posed, by Bricard (1896). Most likely Hilbert simply did not know of Bricard's work, but Dehn did. He mentions Bricard's paper, and another by Sforza[12] from 1897 and says that they both gave examples of polyhedra which cannot be turned into each other by cutting and

---

[11] For this work and its later developments, which pose and solve the problem in any number of dimensions, see Boltianskii's fine treatment (1978).

[12] Which neither I nor Boltianskii have seen.

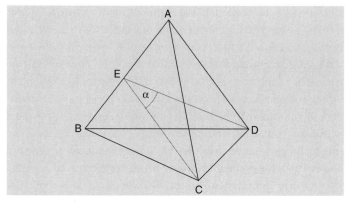

Figure 3

sticking. Indeed, Bricard is quite clear that the relation that he had discovered which must hold between two polyhedra obtained in this fashion, will not hold in general between two polyhedra of equal volume. He wrote (p. 333): 'The fact is almost obvious. One can convince oneself if one wishes by examining a cube and a regular tetrahedron of the same volume.' The honour might seem therefore to belong to Bricard, but his argument is indeed insufficient.[13]

The Padoa affair

Hilbert's 2nd Problem was one of the few that generated comment in Paris. The Italian mathematician Peano observed that more had been done that perhaps Hilbert had seen fit to mention. Giuseppe Peano had started out his mathematical life as an analyst, a branch of mathematics that calls for a high degree of precision. Intuition, especially of a geometric kind, may be one way to get started, but it had become a blunt instrument, capable of leading analysts astray. Peano was adept at discovering and proving counter-intuitive results (such as his function that maps a line continuously onto every point in a square), and it inspired in him a distrust of geometry, and indeed of natural language. He took to thinking

---

[13] It did not consider a sufficiently general decomposition, see Boltianskii (1978) p. 120.

through how mathematics should be written, what symbols it should use, what its basic concepts were. In the 1890s he set about producing a journal of his own that would publish papers in a stripped-down, logical language that he believed would be less misleading than even simple Italian. He also became caught up in movements for an international auxiliary language, of which Esperanto is now the best known. Peano devised his own, which was quite successful for a time. It was in 1900 in Paris, at the International Congress of Philosophers, that Russell met Peano and changed the entire direction of his research.[14]

Peano was therefore intimately aware of issues in the foundations of mathematics. He had attracted a number of followers, and they had written extensively on geometry, from a highly abstract point of view. The discovery of non-Euclidean geometry had forced mathematicians to examine where their deductions had depended on 'knowing' properties of figures without saying so explicitly. Unanalysed assumptions were plainly a possible source of error. The first to take geometry apart with a view to leaving it with no unstated assumptions was Moritz Pasch in 1882.[15] He started with the undefined or primitive concept of the straight line segment between two points, and wrote down all the properties of segments he felt necessary to assume without proof. These he said were immediately grounded in observation, and at this point he cited the authority of Helmholtz, the leading German scientist of the day.[16] He then deduced results from these initial assumptions, of which the first is 'there is always a unique segment joining any two points', arguing that the mathematician should reason logically and without further appeal to sense perceptions. The rest of the book is devoted to showing that that can be done, laying down explicit assumptions when needed.

The Italians around Peano took to this enterprise, but they saw no reason to abstract from intuition in order to produce theories.

---

[14] For an account of this famous incident, see Rodriguez-Consuegra (1991), pp. 135–136.

[15] Pasch (1882)

[16] Helmholtz (1870).

Box 4.1 Hilbert's space-filling curve

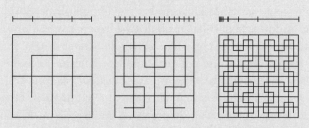

In 1891, inspired by Peano's much more formal example, Hilbert constructed an example of a curve that passed through every point of a square by a series of successive approximations. At each stage every square is divided into four equal smaller squares and the curve replaced as indicated.

In this spirit, Peano noted that the plane case of Desargues' theorem seemed to need assumptions that were not needed in three or more dimensions (see Box 4.2). The Italians produced geometries with fewer assumptions and odder properties, until Mario Pieri completely abandoned any intention of formalising what is given in experience. Instead, as he wrote in 1895, he treated projective geometry 'in a purely deductive and abstract manner, . . ., independent of any physical interpretation of the premises'. Primitive terms, such as line segments, 'can be given any significance whatever, provided they are in harmony with the postulates which will be successively introduced.'[17]

When Hilbert stood up in Paris and alluded to what he had published on the foundations of geometry the year before, it was natural for Peano to feel that Hilbert was slighting Italian work. Hilbert had barely mentioned their achievements, and when he did he gave pride of place to someone outside Peano's group, Veronese. Historians have speculated about what was going on, whether

---

[17] Quoted in Bottazzini (1988) p. 276

there was an element of plagiarism (there is no evidence of that), or if Hilbert was simply unable to read Italian. The truth is likely to be more mundane. While Hilbert could surely read mathematical Italian, as any one can who has studied Latin at school, he was not a particularly well-read mathematician. He preferred to pick up ideas and supply the proof himself, as many mathematicians do. It is much more likely that he simply never bothered to read people like Pieri. Blumenthal tells a nice story about coming to see Hilbert, who was then his PhD supervisor, in some distress because he had found that some of his best discoveries were in fact already in the published literature. Hilbert merely replied 'Why do you know so much literature?'[18]

Still, there is no excuse for ignorance, and Hilbert's neglect of their work seems to have spilled over into the hostile tone of Padoa's response to the second Problem in 1902. Padoa was another member of the group around Peano. He gave a paper at the International Congress of Philosophers on a set of postulates for algebra, and another at the International Congress of Mathematicians on definitions for Euclidean geometry.[19] Now, stung by Hilbert's refusal to engage with his ideas in the intervening two years he went into print arguing that Hilbert's thinking was flawed.[20] Hilbert wanted axioms for arithmetic. Very well, such axioms must be shown to be consistent. To do that, said Padoa, one must exhibit a set of objects that exist and satisfy the axioms. Such a collection would presumably be the natural numbers themselves, so there was a risk of a vicious circle. Hilbert seemed to be calling for a proof of consistency without exhibiting objects, and that, Padoa implied, was absurd. Hilbert does not seem to have replied, and as we shall see there is a way out, but it is hard to believe that either he or Padoa could have suspected it then.

---

[18] Blumenthal (1922) p. 72.

[19] Padoa, (1900) and (1902) respectively.

[20] Padoa (1903).

Hilbert among the philosophers

Hilbert had already been in trouble with some of his views in his *Grundlagen der Geometrie*, and may have felt that it was better not to get drawn into discussions in print. The trouble was caused by his idea, not explained in the first edition but placed in the second, that an axiom system could be what he called complete. By this he meant that there was a system of objects obeying the axioms and there was no larger system of objects also obeying the axioms. He knew it was a different kind of axiom than the others. They described what you could do with the objects; this was a statement about the nature of the axiom system, intended to allow him to claim that the objects really existed. Minkowski had observed this in connection with Hilbert's views on arithmetic, set out in his lecture 'On the Number Concept' at the DMV meeting in Munich, and he wrote in a letter to Hilbert that 'Your existence is as little to be doubted as in your $18 = 17 + 1$ axioms for arithmetic.'[21]

But the axiom drew heavy criticism from Frege, who complained reasonably that it was like doing theology with an axiom that says God exists: 'Axiom 3, there is at least one God'.[22] In other words, completeness axioms cannot be used to resolve questions of existence. Hilbert persisted, and asserted such an axiom for the real numbers, which he announced superseded the need for explicit constructions in the manner of Cantor. He also corresponded with Frege, and got so carried away that he even claimed in a letter to him[23] that 'if the arbitrarily given axioms do not contradict one another with all their consequences, then they are true and the things defined by them exist'. The parallel postulate should have warned Hilbert to be more careful: one can only say that theorems are true of the objects defined by the axiom system, for which a better word is 'proved' (the word Hilbert preferred in Paris), rather than 'true'. But Hilbert refused to get drawn in public, so Frege

[21] Rüdenberg, L. and H. Zassenhaus (1973) p. 116, Minkowski to Hilbert 24 June 1899.

[22] Frege to Hilbert, 6 January 1900, in Frege (1980) p. 46.

[23] Hilbert to Frege, 29 Dec 1899, in Frege (1980), p. 39.

eventually wrote most of his part of the correspondence as an article for the *JDMV*. This probably did Hilbert little damage; Frege's manner was notoriously rude, and it has been argued that Frege's standing in any case remained low until Russell took him up (paradoxically after destroying the basis of Frege's whole analysis of mathematics).

Hilbert also met the former mathematician turned philosopher Husserl when he gave a lecture at Göttingen on 5th November 1901. Husserl went away after some discussion with Hilbert, and filled pages of his Notebook with thoughts about what it could mean to define or create objects by means of an axiom system. Interestingly, it seems as if his immediate point of interest was defining the integers.[24]

The polarity between Hilbert and Frege is a crucial one in modern mathematics. For Frege, existence was primarily a question of what objects there are in the world. Without objects, axiom systems were in his view void. Hilbert was radically of the other opinion: consistency implies existence. Whatever the murkiness of Hilbert's thinking about consistency, his philosophy of mathematics allows the existence of contradictory sets of objects, separately but not simultaneously. One may have Euclidean geometry and non-Euclidean geometry in mathematics, on Hilbert's view, because each has a consistent set of axioms. In fact, mathematicians had proved that if one accepts one of these geometries one must accept the other (they are, in the jargon, relatively consistent). But on Frege's view (most clearly in unpublished material) there is only one world and so only one geometry, and the other one (non-Euclidean geometry, it would seem) is simply meaningless. This shows that Frege had no grasp at all of the relative consistency argument.

Another way to see the distinction is between axiom systems that codify, and axiom systems that create. Codifying what is known is a traditional activity, almost certainly conducted by Euclid, certainly conducted by Pasch for geometry and by Helmholtz for the integers. It is also, as we shall see below, what Hilbert was doing when he produced axioms for branches of physics. But what

---

[24] See Majer (1997).

Hilbert wanted to do, as did the Italians, was to create mathematical objects by giving rules for their use. This was quite a novel approach, and it is not surprising that there was some doubt about how it can legitimately be done.

Axiomatising geometry in USA, responses to *Grundlagen der Geometrie*

One of the main communities of mathematicians that picked up on the idea of axiomatising geometry, and in so doing spread its larger message that axiomatising might be an important way to proceed in general, was the group of young mathematicians around E.H. Moore at Chicago. They already had strong ties to Göttingen and were well placed to run with this idea.[25] Moore himself had attended the ceremony for the unveiling of the Gauss–Weber memorial in 1899, the occasion for which Hilbert wrote his *Grundlagen der Geometrie*, and Moore took the new ideas back to America, where in 1901 he led a seminar on the foundations of geometry, and lectured on Hilbert's work. Hilbert's axioms had already been criticised by Friedrich Schur for their lack of independence, and Moore discovered that although Schur was right, the redundancies were incorrectly identified. His presentation of this discovery in class drew the gifted young mathematician Oswald Veblen into the study of geometry, and 18 months later Veblen presented his PhD thesis, in which he based Euclidean geometry on 12 axioms based on the notions of point and order. These were published.[26] He then went on to spend many years on various aspects of the foundations of geometry, and in due course became one of the leading American geometers and a major influence on the Institute for Advanced Study at Princeton.

Veblen's axiomatic treatment of the foundations of geometry is notable for its rigour, and remarkable for the thoroughness of the literature it cites. He himself identified it as belonging more to the

---

[25] See Parshall and Rowe (1994).
[26] Veblen (1905).

Box 4.2 Moulton's counter-example to Desargues' theorem in the plane

Moulton in 1902 divided the familiar plane into two by a line called the Axis. He then defined lines in a novel way. A line is either a conventional line parallel to the Axis, or a conventional line crossing the Axis but sloping to the left, or a new kind of line obtained from a line crossing the Axis and sloping to the right. If such a line makes an acute angle, β, with the Axis at a point P, it is replaced by a broken line composed of the half of the original line below the Axis and a line segment above the Axis making an angle α at P with the Axis, where $\tan \alpha = \frac{1}{2} \tan \beta$.

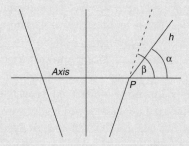

The usual axioms of projective geometry are satisfied, for example two lines meet in at most one point, but in this geometry, Desargues' theorem is false. The real force of this result is that it shows that Moulton's plane cannot be embedded in any three-dimensional space (if it could, Desargues' theorem could be proved). Hilbert used this example in later editions of his *Foundations of Geometry*.

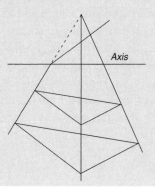

tradition of Pasch and Peano than that of Hilbert and Pieri, but the significance it has for us is that it belongs, like much American work of the period, to the reception of Hilbert's ideas as being axiomatic in spirit. The American school of postulation theorists, as they have been called,[27] did indeed write down axiom systems for many things, not only geometries but groups and even logic. The point of the enterprise is two-fold. One aim is always to spell out the independence of the axioms and their role in establishing the main features of the theory. Hilbert himself was to emphasise this point repeatedly. If an axiom system is used in a given context it may be that one of the axioms comes to seem doubtful, and alternatives are proposed. A properly constructed axiom system allows the researcher to see very quickly what the consequences of such changes might be, whereas a mere list of working assumptions does not (with possibly catastrophic results).

The second use for an axiom system is to characterise its objects. This provides a clean description of the objects concerned, which might otherwise be difficult to understand. It enables the key features of a problem stand out clearly, and it allows mathematicians to recognise that two superficially different problems are 'really' the same, and that methods which work well with one problem can therefore succeed with the other.

It has often been alleged that the axiomatic approach to mathematics licenses mathematicians to write down any set of axioms, satisfy themselves that they are not self-contradictory, and then proceed to deduce consequences of the objects they define. While there is some truth in this allegation, this was not Hilbert's view, nor that of his leading German and American followers. They insisted that the body of knowledge being axiomatised was of independent significance. Hilbert's examples in his list of 23 problems had to do with set theory, arithmetic, Euclidean and related geometries, Lie groups, and physics. The strength with which he and others held this view has sometimes been forgotten, or dismissed as mere aesthetics, but it was far from that, and when it was used most imaginatively it was a spectacular success, as the exam-

---

[27] Scanlon (1991).

ple of Emmy Noether's creation of abstract (structural or modern) algebra shows.

Another of the American mathematicians most involved in responding to Hilbert's ideas on geometry was also one of its great originals, hailed by R.L. Wilder in his account of him as both 'probably one of the most influential American mathematicians of the first half of the twentieth century' and as one 'possessed by dogmatic prejudices' (including complete certainty in the truth of the Axiom of Choice).[28] This was Robert Lee Moore.[29] A Texan by birth, he went to the University of Texas in 1898, where he came under the spell of G.B. Halsted, the man responsible for many of the translations into English of the key texts on non-Euclidean geometry. Halsted put Moore onto Hilbert's *Grundlagen der Geometrie*, and asked him to check if all the axioms listed there were independent. Moore found that they were not, and this led to his first published paper (the same discovery had been made by E.H. Moore earlier, but he gracefully ceded the publication to the younger man). From 1903 to 1905 he was at the University of Chicago, which as we have already remarked was the undoubted centre of mathematical research in America at the time and intimately connected to Göttingen. There he met Veblen, who had just written his thesis on the axiomatic foundations of geometry. But whereas Veblen's career took off rapidly, Moore had to spend over a decade of insecurity before returning to Texas, where he wrote prolifically and pioneered a teaching method which avoided text books of all kinds and required students to work through a structured set of results. The so-called Moore method has its passionate advocates to this day, although others feel that it succeeds in producing creative mathematicians at the price of producing narrowly educated ones.

R.L. Moore wrote on many aspects of point-set topology, and the

---

[28] R.L. Wilder (*AMS History*, 3, 265). The Axiom of Choice is the claim that it is possible to select simultaneously an element from each of an infinite collection of non-empty sets. This intuitively plausible claim has some surprisingly counter-intuitive consequences.

[29] It seems E.H. and R.L. Moore may have been seventh cousins.

papers we shall consider date from his earliest period of work, when the influence of Hilbert was particularly strong. He wrote, but never published, an attempt to characterise the positive integers and their arithmetic (reminiscent of Hilbert's 2nd Problem). He did publish a set of axioms for plane Euclidean geometry. This led him to think about topology, and as his fallow period came to an end he wrote on the foundations of plane geometry from a more topological point of view.[30]

He began the paper with Poincaré's criticism of the *Grundlagen der Geometrie*. This was that while the hypotheses about groups that Hilbert made are more general and satisfactory than Lie's (and closer to the 5th Problem, we could add) the assumption about the space for the geometry is still restrictive. To be precise, it is assumed to be a number plane, even though, Poincaré conceded, it is analysed more carefully than anyone had before. Moore set himself the task of defining a geometry on the basis of the primitive or undefined notions of point, region, and motion. It is worth noting that the belief that geometry involves us knowing about the whole space and not just pieces of it is a tenacious one. Euclid had it, as his use of the parallel postulate attests, and so did all nineteenth century geometers up to and including Klein. Moore and Veblen were among the first to restrict assumptions to purely local ones, a philosophy of geometry that only became widespread after Einstein's theory of general relativity, which is a firmly local theory.

Moore's 12 axioms governing the use of his undefined notions were naturally highly topological in nature. They drew on his paper[31] in which he had given an axiomatic foundation for plane topology, and here they formalise the idea that space is a homogeneous collection of points which can be moved around, and the neighbourhood of each point looks the same. The mental picture is a uniform fog. Moore showed that some of the axioms force a region to be very much like a disc, without holes and having a boundary like a circle. By the end of the paper he has recaptured, from his more general starting point, the ideas that Hilbert had

[30] R.L. Moore (1919).

[31] R.L. Moore (1916).

encountered, proving that curves defined by motions and which ought to be circles actually are circles. It follows that the underlying space of points actually is a number plane, and that the underlying geometry is either Euclidean or non-Euclidean.

Hilbert and Brouwer on Hilbert's Problem 5

Hilbert's fifth Problem was taken up by the young Dutch mathematician L.E.J. Brouwer, who presented his preliminary findings at the International Congress of Mathematicians in Rome in 1908. He then gave the details of his work in two long papers in the *Mathematische Annalen*, thus firmly promoting his name in association with the solution of what he took to be an important problem. Moreover, in 1909 Brouwer, by then 28, met Hilbert in person for the first time, when Hilbert spent a summer holiday at Scheveningen (now close to the site of the main airport for the Netherlands, but then a seaside resort).

From van Dalen's recent account of Brouwer's early life it seems that the meeting went well for both men.[32] Brouwer took the opportunity to discuss with Hilbert his ideas on the foundations of mathematics and to disagree with Hilbert's address to the Heidelberg ICM of 1904, and the discussion was stimulating enough for Hilbert to seek out Brouwer in his summer cottage. Brouwer was a passionate man with a personality oddly akin to the haunted souls of contemporary expressionist music, and he was enraptured by Hilbert. He described their meetings as shining 'a beautiful new ray through my life'.[33] As for the work on Lie groups, Brouwer had recently discovered that the first drafts of his second paper for the *Annalen* were flawed by an incautious reliance on a paper by Schoenflies on topology, and was pleased that he had been able to understand the topology correctly. This had driven him back to consult Hilbert's paper of 1902 on the Lie group of the Euclidean plane, and in October 1909 he wrote to Hilbert explaining what needed to be done to make Hilbert's paper truly

---

[32] van Dalen (1999).

[33] van Dalen 1999, p. 128.

rigorous. (It is not known how Hilbert responded to these comments.)

Brouwer's work, like Hilbert's, was far from being general, and was restricted to groups that act on either the line or the plane. It was necessarily technical, but it was also a significant step in the development of the nascent subject of topology. The way forward had been outlined by Hilbert himself. Hilbert had attempted to characterise the groups that can arise as transformation groups acting on the Euclidean plane. He therefore defined a plane and a transformation group. He defined a plane as a space modelled on the familiar number plane and such that every point has a neighbourhood which is the interior of a simple closed curve (See Figure 4).

A simple closed curve is a curve which is a continuous image of the circle and which does not cross itself. Such curves are called Jordan curves, after the French mathematician Camille Jordan who had been the first to draw attention to their fundamental nature. Hilbert defined a motion as a continuous invertible transformation of the plane that does not alter the orientation of the boundary of any neighbourhood, and he required that the set of all motions forms a group. He also required that, given two points A and M, there are infinitely many different motions that fix M and have distinct images for A. The totality of positions for the point A he called a true circle, so his axiom implied that a true circle contains infinitely many points.[34] The bulk of the paper was devoted to showing that true circles are Jordan curves. Once this was done the only possible geometries turned out to be plane Euclidean geometry and two-dimensional non-Euclidean geometry. Since the transformations were continuous but had never been required to be analytic, this established that at least in these cases Lie's theory of transformation groups did not need to presuppose analyticity.

The topology turned out to be remarkably tricky. The problem is that a curve can be a very unexpected object, and by 1900 mathe-

---

[34] Hilbert also stipulated that if there is a sequence of transformations taking a triple of points arbitrarily close to a triple ABC to a triple of points arbitrarily close to a triple $A'B'C'$ then there is a transformation taking ABC to $A'B'C'$.

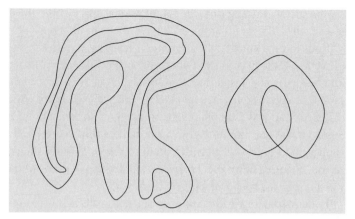

Figure 4 (a) A simple closed curve. (b) Not a simple closed curve.

maticians were shocking themselves with the properties curves can have. Mathematicians found curves that are not differentiable anywhere (they never have a tangent). They found curves that map a line onto an entire square; as we have seen, Hilbert himself gave a simple construction for such a space-filling curve. The space-filling curve necessarily crosses itself, so it is not a closed Jordan curve, but the American mathematician Osgood gave examples of Jordan curves of finite area, which are equally counter-intuitive.[35]

Perhaps the crucial property of a closed Jordan curve in the plane is that it separates the plane into two parts, the 'inside' and the 'outside', in the sense that a line cannot be drawn from one region to the other that does not cross the curve. Another key property is that the curve divides the plane into two regions, one (the 'inside') topologically equivalent to the open disc and the other topologically equivalent to the open disc with a point removed. Jordan and others had indicated how these claims might be proved, and indeed they are true, but in 1908 there was still no valid proof. On the other hand, it was one of Schoenflies's mistakes to make analogous claims about spheres in three-space. The much

---

[35] See Sagan (1994) for many examples of this kind.

more difficult topology of three-dimensional space is the main reason why Hilbert and Brouwer stopped at two dimensions.

It was to turn out that the continuous image of a sphere in the plane that does not cross itself does separate Euclidean three-space into two parts, the 'inside' and the 'outside', in the sense that a line cannot be drawn from one region to the other that does not cross the curve. But it is false that of the two regions so formed one is topologically equivalent to the open ball and the other topologically equivalent to the open ball with a point removed. This was shown most dramatically by the American mathematician Alexander when he defined his horned sphere in 1924.

Brouwer set himself the task of enumerating all the Lie groups that could arise from any kind of two-dimensional geometry, and after resolving a number of deep problems in point-set topology he succeeded. But his work did not convince everyone, and most notably it failed to convince a leading expert of the old school. This was Friedrich Engel, who had indeed worked with Lie for many years (he had originally been sent out by Lie's friend Klein, who rightly suspected that Lie needed a co-author if he was ever to write his ideas out at length and intelligibly). He was sent Lie's two papers in the *Mathematische Annalen* to review for the abstracting journal of the day, and his puzzled reviews sparked a correspondence between the two in which the older man confessed that he did not fully grasp Brouwer's set-theoretic formulations,[36] and in second review he continued to lament: 'I am still of the opinion that anyone who is not a set theorist incarnate, will find the general hypotheses of § 1 not expressed sufficiently clearly.'[37] ' I must say, he wrote to Brouwer, "the expressions of the system sound obscure to uncircumcised ears".'[38] This splendid grumble marks the shift from analysis to topology, and one generation to another.

---

[36] *Fortschritte der Mathematik*, **40**, 1909/1912, p. 194.

[37] *Fortschritte der Mathematik*, **41**, 1910/1913, p. 181.

[38] Van Dalen (1999) p. 131. For the whole matter see pp. 126–133.

Box 4.3 The Alexander horned sphere

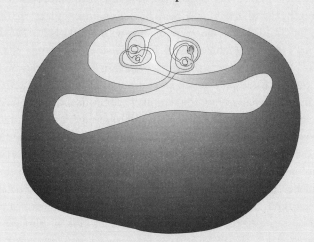

It is constructed by drawing out two 'horns' from the sphere and bringing them close together. From the tip of each horn, draw out two more, and twist them round as shown. Now imagine this construction has been done infinitely often; the result is the Alexander horned sphere. It is easy to believe (but not so easy to show) that a loop passed around the first horn cannot be removed, which shows that the 'outside' of the horned sphere is not topologically equivalent to an open ball with a point removed.

[From J.G. Hocking and G.S. Young, *Topology*, p. 176]

Hilbert Problems on pure and applied mathematics

One of the exciting discoveries made by historians of mathematics in recent years has been the extent of Hilbert's interest in physics. In truth, the evidence has been sitting almost on the surface all the time, and only its relative absence for the three volumes of Hilbert's *Collected Works* and some rather dismissive remarks by Hermann

Weyl in his obituary of Hilbert seem to have misled people. Hilbert's interest was the logical outcome of his education at Königsberg, where it occupied a distinguished place in the syllabus, and of his close friendship with Minkowski. It was through Minkowski that, in 1894, Hilbert made a close study of the ideas of Heinrich Hertz and first principles in physics. Hertz is famous for his experimental creation and detection of electromagnetic waves, and until his untimely death from leukaemia at the age of 37 he was regarded as the bright young man of German physics. Leo Corry, who has been at the forefront of revising our picture of Hilbert, describes the effect of Hertz's ideas on Hilbert as seeming 'to have provided a final significant catalyst towards the wholehearted adoption of the axiomatic perspective for geometry, while simultaneously establishing, in Hilbert's view, a direct connection between the latter and the axiomatization of physics in general.'[39]

What Corry has brought to light is the great number of lectures courses that Hilbert gave on various branches of physics at Göttingen; notes of many of the courses survive in Göttingen.[40] The first course, on mechanics, was given in 1898 (the last, on mathematical models of quantum theory, in 1926/27). When he next lectured on mechanics, in 1905, he offered an axiomatic account. The physical concept of force is described as a vector, axioms describe the behaviour of vectors, and prescribe that vector arithmetic is continuous. Hilbert felt very strongly that continuity was built into physics because it was verified by experiment. Further axioms described statics, and then space, time, and motion. So far, so routine, except that it has to be done well and to effect. Hilbert, for example, deduced Newton's laws of motion from his formulation of statics, and while his account of time is murky he could with justice say that his derivation of the laws of motion surpassed Newton's. More substantial is the introduction of the first truly fundamental piece of physics: the law of conservation of energy. This proves to be characteristic of Hilbert's

---

[39] Corry (1997), p. 105

[40] For a list, see Corry (1999).

axiomatic approach to physics. The nub of the matter, for him, was to express a law that plays a fundamental role in developing the theory. Such rules, given the status of axioms, need not be self-evident. Hitherto, the usual idea of an axiom was that it was an undemonstrable but obvious truth, required because reductionist explanations must stop somewhere. Hilbert preferred axioms that played a decisive role in the theory they underpinned, and the ones he chose were often couched in familiar terms, such as conservation laws and variational principles.

Hilbert also compared his axiomatic account of physics to those of Hertz and Boltzmann[41] praising Hertz for the wonderful structure of mechanics he presented which gave a complete system of axioms and definitions that attained a Euclidean ideal. But he nonetheless disagreed with Hertz's proposal to banish force from physics.[42]

Hilbert included thermodynamics in his course. This treats heat as a form of motion, and presents a number of intellectual difficulties. The concept of entropy, which measures the disorder of a physical system, is almost notorious in this connection. Max Born, who had studied in Göttingen, studied Gibbs' book on the subject while in Cambridge in 1907 and proposed to a friend of his that the central results of that book called out for a proper mathematics formulation that he, Born, could not find. In so doing he was echoing Hilbert's views from lectures he had heard in 1905. The friend was Constantin Carathéodory, another Göttingen graduate, and in 1909 he published his own axiomatization of thermodynamics. It seems, however, that whatever its merits, Carathéodory's axiomatic presentation did not catch on. Born tried to revive it in 1921, but while this time he attracted the Cambridge physicist R.H. Fowler, he otherwise failed. Perhaps, as Born suggested, Carathéodory's presentation simply was too abstract for physicists, but it was a significant contribution to Hilbert's 6th Problem all the same.

Hilbert was also interested in probabilistic theories, such as the

---

[41] See Corry (1997).

[42] See Lützen (1999).

kinetic theory of gases, which successfully finessed by statements about averages, the need to write down several trillions of differential equations to describe the motion of every particle in a gas. This was an area where Hilbert expressed a worry to Frege back in 1899 that physicists wrote self-contradictory nonsense. As he put it:

> After a concept has been fixed completely and unequivocally, it is on my view completely illicit and illogical to add an axiom—a mistake made very frequently, especially by physicists. By setting up one new axiom after another in the course of their investigations, without confronting them with the assumptions they made earlier, and without showing that they do not contradict a fact that follows from the axioms they set up earlier, physicists often allow sheer nonsense to appear in their investigations.[43]

In particular, it was not clear when the probabilistic elements of the theory were introduced in order to get (qualitative or quantitative results) and when they followed from the theory as it stood (or from some simpler theory). One of the successes of Hilbert's theory of integral equations in 1912 was his derivation of what are called Boltzmann's equations, and he also had three doctoral students working on the subject in 1913–14.

In 1915 Hilbert's interests swung decisively onto what Einstein was doing in the general theory of relativity. Corry has recently discovered the first version of Hilbert's paper on the subject, which shows clearly that in fact Einstein was well ahead, contrary to every prevailing history of the topic. Moreover, as Rowe has shown[44] Hilbert was in fact trying to do something rather different. Still, it is very clear that Hilbert, and indeed Klein and the young Emmy Noether, were highly serious in their attempts to provide genuine mathematics support for the best of current physics. And even if, as Rowe argues, Hilbert's overly axiomatic approach is in some ways the weakest it shows eloquently how importantly he regarded his 6th Problem.

Hilbert's 6th Problem was not, therefore, a mere gesture towards

---

[43] Hilbert to Frege, December 29, 1899, quoted in Corry (1999).

[44] See Rowe (1999) and forthcoming papers.

a topic mathematicians are supposed to find important, especially on public occasions. Physics was a genuine, life-long interest of his, and the axiomatic approach his specific contribution. He saw geometry as the transitional case, because geometry belongs to mathematics and yet has an empirical aspect. Beyond it stood mechanics, which should yield to the axiomatic treatment once it is more fully understood. Beyond mechanics lay the other branches of physics, which the mathematician could hope to sort out as and when the physicists reached a consensus. Writing in 1922, Max Born observed that the Göttingen tradition of Gauss, Wilhelm Weber, and more recently Klein had sought to develop mathematics and physics not side by side but together, and to put mathematics at the service of science. From this perspective, Hilbert had been less interested in the practice and more in the underlying principles. Although he did not say so, he may well have felt that this removed Hilbert from the fray, and the logical and mathematical arguments were too desiccated for the physicist. Perhaps, too, the research physicist loses interest in a subject as consensus emerges. It might be that Hilbert would have done better to claim that axiomatic mathematics can resolve contentious issues in physics. That, however, would have exposed him to the charge of arrogance, doing physics without experiments. The separation of mathematics had become quite broad by 1900, broader even than Hilbert could manage.

Problems 19, 20, and 23

These problems form a coherent topic, and one that was to grow dramatically in the twentieth century. Hilbert's interest in elliptic partial differential equations and variational problems are closely connected, as his example of the Dirichlet problem shows. It will be best to start with some simple but important examples. In that way we can seek to isolate Hilbert's perspective on these questions from others that were current at the time, and to disentangle the threads that lead back to his problems.

The shape of a vibrating string—say a guitar or violin string—determines the note it makes. A good description of the motion, which explains the basic features of music on such instruments, supposes that the string is initially straight, but when plucked moves in a plane. Each point on the string moves sideways, perpendicular to the rest position of the string, in a way determined in part by the length of, and tension in, the string and in part by the shape it is put into by the player. As the string moves, the position of a point P on the string varies in the perpendicular direction, and so at each moment in time each point of the string has two coordinates, one giving its position along the string and the other its departure from the rest position.

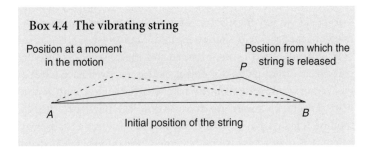

Box 4.4 The vibrating string

Position at a moment in the motion

Position from which the string is released

P

A

B

Initial position of the string

The string is released from the position shown in the diagram. Throughout the subsequent motion, it retains a v-shape which runs up and down the string. The frequency of this motion determines the pitch of the note, while the initial displacement of the point P determines the amplitude or volume of the note. Each point on the string moves like a pendulum bob, oscillating on either side of the rest position.

In mathematical terms, the motion of a vibrating string is described by an equation called the wave equation, due to the eighteenth century French mathematician Jean D'Alembert:

$$\frac{\partial^2 y}{\partial t^2} = c^2 \frac{\partial^2 y}{\partial x^2}.$$

The second derivatives tell us that acceleration is the key physical process. It is produced by the forces in the string that try to push it back into the rest position. The precise solution (and hence the note) depends on the initial conditions, but these are likely to be complicated. They are, in fact, the initial shape of the string, which is entirely arbitrary (forced vibrations are no simpler, which is why listening to a beginner play the violin can be such an excruciating experience).

The study of gravity, and the gravitational attraction of irregularly shaped bodies in particular, led mathematicians at the start of the nineteenth century to a second differential equation. It is deceptively like the wave equation, but its properties are very different. It is the potential equation:

$$\frac{\partial^2 u}{\partial x^2} + \frac{\partial^2 u}{\partial y^2} = 0.$$

The change of variables is intended to highlight the nature of the problem. The unknown function, u, depends on two variables, x and y, which are both now thought of as position coordinates (there is no variation with time). As with the wave equation, there will be a large family of solutions to this differential equation, and in any given case the relevant one is picked out by some further requirement.

The problems start with the further requirements. A traditional setting in which this problem arises asks for the equation of a surface defined by a solution of the differential equation and spanning a given loop (this is the famous Dirichlet problem).

Each point on the surface is a certain height, u, above the (x, y)-plane, and the height is given by the solution of the differential equation that also passes through the loop. It is reasonably clear that there is always such a surface, but proving it is much harder. The first indication that this is so is that the solutions are far from arbitrary. Functions $u(x, y)$ that satisfy this equation are said to be harmonic. They have a property which came to be seen as more and more unusual as the nineteenth century wore on: they can be differentiated infinitely often (they are said to be 'regular' for this reason). The claim that a boundary curve—which might not be smooth at all—spans a surface which is extremely smooth at every interior point is not so very plausible. Yet the physics of it is entirely seductive. When the problem is seen as one with a physical interpretation, the existence of a unique solution for any boundary curve is quite obvious. Physicists and mathematicians were divided on the issue, as mathematicians found the problem more and more subtle and the existence of solutions harder and harder to establish, while physicists had their intuitions supported by experiments, and were at times disdainful of mathematical pedantry.

Moreover, it is a general rule in differential equation theory that the solution functions are harder to understand than the data that defines the equation. So the regularity of the solutions to the potential equation is also remarkable on these grounds.

We are now in a position to review Hilbert's 19th Problem. First he drew attention to the strange regularity of solutions to the potential equation, and to certain other equations (the case of the minimal surface equation is considered below). Then he singled out a way of looking at these equations which made them seem as if they belonged in a family: they are what he called 'Lagrangian' equations of a 'regular variation' problem. Then he conjectured that for this class of equations, the solutions will always be regular. So while Hilbert did look for a class of differential equations to which the potential equation belonged, he drew the line tightly enough to exclude many other types of equation, notably the wave equation.

In his 20th Problem, he shifted focus to the very existence of solutions. He noted that while the Dirichlet problem had been solved for a large class of boundary curves the method did not seem to extend to similar problems, and he asked for a general method. Then he took on the tension between mathematicians and physicists and asked if there would always be solutions if the boundary conditions were not quite arbitrary. Could one delineate a class of boundary conditions which were general enough to cover the physical situations and yet narrow enough to finesse the mathematical difficulties? If pressed, Hilbert admitted, one might retreat a little on the idea of what was meant by a solution. In short, Hilbert pointed to a large class of equations, containing several that arise in various physical problems, and asked for conditions under which they have particularly regular solutions, and for general theorems guaranteeing the existence of solutions, perhaps in some attenuated sense to be made precise by future research.

For the case of the potential equation, Hilbert wanted a general setting which was any second order elliptic partial differential equation,[45] and he claimed that it should be possible to show that if z was a reasonably well behaved function of x and y, then all these equations had solutions that were analytic. Partial results in this direction had indeed been obtained by the French mathematician Émile Picard in 1895. He showed that if the potential equation has some simple terms involving only the first derivatives added to it, all the solutions remain analytic.[46] In 1902 a student of Hilbert's called Lutkemeyer, and independently Holmgren the next year, showed that the theorem is true for equations where an entirely general first order term is added. This establishes that, in particular, all surfaces of constant positive curvature are given by analytic expressions.

The final step to the solution of Hilbert's 19th problem was then taken by the young Russian mathematician Serge Bernstein in 1904, when he showed that the solutions of any well-behaved elliptic partial differential equation are analytic if they satisfy some simple requirements on their first three derivatives.[47] Bernstein was quite explicit that he had solved one of the Hilbert problems, and although he had been based in Paris, he spent three semesters in Göttingen between 1902 and 1904, and submitted his paper to the *Mathematische Annalen* (a journal edited by Hilbert) from there. He was only 23 at the time.[48]

---

[45] A partial differential equation of the form

$$A\ \frac{\partial^2 z}{\partial x^2}\ +B\ \frac{\partial^2 z}{\partial x \partial y}\ +C\ \frac{\partial^2 z}{\partial y^2}$$

plus first-order terms $= 0$ is elliptic if the functions A, B, and C satisfy $B^2 - 4AC < 0$. This is by analogy with the equation for a conic section: $Ax^2 + Bxy + Cy^2 + Dx + Ey + F = 0$ defines an ellipse if and only if the numbers A, B, and C satisfy $B^2 - 4AC < 0$. The interesting thing is not that this analogy can be set up, but that it makes good sense.

[46] Picard (1895).

[47] Bernstein's equation looks like

$$F\left(x,y,z,\ \frac{\partial z}{\partial x},\ \frac{\partial z}{\partial y},\ \frac{\partial^2 z}{\partial x^2}\ ,\ \frac{\partial^2 z}{\partial x \partial y}\ ,\ \frac{\partial^2 z}{\partial y^2} \right) = 0,$$

where the function F is analytic.

[48] I am indebted to K.-G. Steffens for this information.

With the wisdom of youth on his side, and some of Hilbert's contagious optimism, Bernstein painted a rich setting for his work on Hilbert's 19th problem. His paper begins: 'All mathematicians and physicists seem to agree today that the domain of application of mathematics has no other bounds than the limits of our own knowledge itself.' He then asserted that among the best developed modern theories that of analytic functions holds pride of place, indeed its importance was increasing day by day. It acquired such importance, he said, in part because a large class of partial differential equations has analytic solutions—thus connecting his paper with some of the most powerful currents in contemporary mathematics. As for his own 'modest researches', he thanked Picard for creating the appropriate mathematical methods, and Hilbert most particularly for 'having personally recommended this interesting subject to me'.

Bernstein showed that his method was completely general, and worked for all second-order partial differential equations in two variables of Hilbert's type. Because it relied on estimates of the size of the first three derivatives of a solution which were to be known in advance, he called this the method of *a priori* estimates, and the name has stuck. It has been used ever since to study elliptic partial differential equations whenever such estimates can be found, and considerable skill has been developed in finding such things. Still, as Enrico Bombieri said in his address to the International Congress of Mathematicians in 1974, where he was awarded a Fields Medal for his work in this area, 'the search for such estimates in the non-linear case is still today more of an art than a method.'[49]

Problem 23

Hilbert ended his list of Problems with a long account of what was to be done with the calculus of variations. After the brisk presentation of the previous four problems, the lengthy disquisition here suggests not only that Hilbert found the topic very important but

---

[49] It should also be said that there have been several profound shifts in the formulation of this problem, and it would easily take a book just to describe what has been done on the topic since Bernstein's original work.

Box 4.5 Bernstein's theorem

Bernstein took the specific example of the partial differential equation

$$z \frac{\partial^2 z}{\partial x^2} + \frac{\partial^2 z}{\partial y^2} = 0. \quad (*)$$

He supposed equation $(*)$ had a solution $z = z(x, y)$ on a domain S containing the origin, at which $z(0,0) = 1$. He also assumed that the solution and its first three derivatives are bounded inside and on a circle C with centre the origin and lying entirely inside S.

Bernstein then considered the sequence of equations

$$\frac{\partial^2 u_0}{\partial x^2} + \frac{\partial^2 u_0}{\partial y^2} = 0,$$

$$\frac{\partial^2 u_1}{\partial x^2} + \frac{\partial^2 u_1}{\partial y^2} = (1 - u_0) \frac{\partial^2 u_0}{\partial x^2},$$

$$\frac{\partial^2 u_n}{\partial x^2} + \frac{\partial^2 u_n}{\partial y^2} = (1 - u_{n-1}) \frac{\partial^2 u_{n-1}}{\partial x^2},$$

The function u_0 is the harmonic function which agrees with the solution function z to equation $(*)$ on the circle C. The successive functions $u_1, u_2, \ldots, u_n, \ldots$ get closer and closer to a solution of the equation $(*)$. Bernstein wrote $u_n = u_0 + v_1 + \cdots + v_n$, where each v_n vanishes on the circle C. The sequence $u_1, u_2, \ldots, u_n, \ldots$ converges to a function u provided that sum $u_n = u_0 + v_1 + \cdots + v_n$ converges uniformly, which it does because of his assumptions about the function z. Moreover, the function u is a solution of equation $(*)$.

The solution, u, of equation $(*)$ agrees with the solution z on the circle C and is analytic. He then showed that the solutions u and z agree inside C, and therefore concluded that the solution z was itself analytic. Since it had been assumed to be only three times differentiable, this was a remarkable improvement.

that he was beginning to have quite precise ideas about what could be done to advance it. Such was indeed the case.

The topic is a fundamental one, and is directly connected with the problem of the curve of shortest descent that he had emphasised at the start of his address. Many problems in physics and in mathematics itself ask for a function, a curve, or a surface satisfying an extremal property. For example, Fermat's principle in optics—to take a different example—states that the path taken by a ray of light joining two points is the one that the light takes the least time to travel. So to find the path by means of this principle one would like to

1. write down an arbitrary path,

2. compute the time light takes to travel that path, and

3. pick a (usually unique) path for which this time is the minimum.

Because the last step cannot be done by scanning a list of all the times (which would be infinite) mathematicians instead provide a characterisation of the path that yields a minimum. (This is akin to the much simpler question of using the calculus to find the minimum or maximum value of a function $y = f(x)$. The extremal values are picked out by the solutions of the equation $f'(x) = 0$.) If the medium through which the light is passing is homogeneous the solution will, of course, be a straight line. If it is air and water, the path will be a broken line, and in a variable medium the light will take various curved paths, which can give rise to mirages.

In the calculus of variations the property is usually presented to the mathematician in a local form (in the above example, the velocity of light at each point in the medium is known, so one knows how long the light takes to travel an 'infinitesimal' distance). The quantity to be minimised is then found by integration (which 'adds up' the elapsed times). So steps (1) and (2) above are usually formulated as:

1. write down an arbitrary path, and

2. compute the integral of a specific function which depends on the arbitrary path.

The third step asks for a way of characterising the path which gives this integral an extremal value, from which the description of the path can be obtained explicitly. It is important to note that just as solutions to the equation $f'(x) = 0$ might be maxima, minima, or neither (e.g. the point x = 0 on the curve $y = x^3$) so too an extremal might maximise the integral, minimise it, or fail to do either.

The whole force behind the problem of the calculus of variations lies in a remarkable disjunction between what mathematicians and physicists believed to lie at the heart of their respective disciplines. The physicists, and perhaps especially the German physicists such as Helmholtz or Kirchhoff, placed enormous reliance on these variational principles. Their multi-volume treatises of physics encapsulated many physical laws in terms of the calculus of variations. The good mathematician around 1900 was, on the other hand, becoming uncomfortably aware that mathematics could not supply the intuitions of the physicist with the security of mathematical proof. Either nature displayed a preference for regularity and an increasingly large domain of mathematics would be forever without application, or nature was not so selective and the physicists had been naïve. To be sure, the evidence suggested that the physicists were right, but the gap was certainly real and opening up. The contribution of the mathematician could not be to close it, or (as it seemed in 1900) to provide applications for more mathematics, but to characterise in some way the useful parts of mathematics so that the proofs mathematicians supplied reflected nature's parsimony.

The mathematical problem lay at the same juncture Hilbert had highlighted in the text of his 5th Paris Problem, namely that critical node where the differentiable and the merely continuous diverge. Differentiable functions have, by and large, the properties naïve intuition suggests, and the more differentiable they are the better behaved they appear, culminating in the analytic functions, which are remarkably well behaved. A differentiable function has a graph with a tangent at each point. It makes sense to talk of the graph being drawn by a moving point (informally, the tip of a pencil). To find the extremal values of a differentiable function one indeed differentiates it and equates the resulting function to zero. But if the function is not differentiable one cannot do that, and other more ad hoc techniques have to be used to find the minimum.

A serious problem is presented by functions which are nowhere differentiable. By 1900, such functions were turning up all over the place. There was indeed a cottage industry turning out functions with a wide variety of counter-intuitive properties, for which the traditional methods of the calculus do not apply.

Physicists in Hilbert's day could afford to dismiss these functions as irrelevant. What, however, was the corresponding situation in the calculus of variations? Hilbert said that although the topic was important, it was, in his opinion, not sufficiently appreciated by mathematicians. That was perhaps because it lacked a good text book (which made the new one by A. Kneser so valuable).[50] Most recently Weierstrass had discussed it at length, but Hilbert's method was new and simpler. He proposed to tackle the difficulties head on, somewhat in the way the differential calculus picks out extremal values of functions by characterising the extremal value among all nearby choices of the variable. He suggested that the calculus of variations determine extremal curves by some kind of limiting argument.

A simple illustration of what can go wrong is provided by this question: find the curve of shortest length joining the points $(-1, 0)$ and $(1, 0)$ in the plane from which the origin has been removed. One can easily draw curves whose lengths exceed 2 by any arbitrarily small amount, but the only curve of that length joining the two points is the segment of the straight line through the origin, and that has been excluded (Box 4.6).

One final paradox about the length of curves indicates another difficulty. Here curves are drawn very close to a curve we wish to measure, in a way that suggests $2 = \sqrt{2}$ (see Box 4.7).

In his published paper, Hilbert sketched a new method which aimed at establishing when an extremal is actually a minimum. Our interest, however, is not in the details of Hilbert's method, but

---

[50] A. Kneser (1900).

Box 4.6 No extremal velocity for as long as possible

Consider the problem of joining the points $(-1, -1)$ and $(1, 1)$ by a differentiable curve, $y = f(t)$, where we are to think of the derivative $\dfrac{dy}{dt}$ as the velocity of the curve. The only condition on the curve is that it should have zero velocity as much of the time as possible. There are many curves for which the velocity is non-zero only during a very small amount of time, a particularly symmetrical one is shown in the diagram.

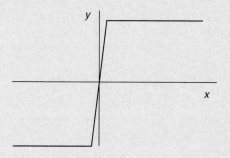

The corners are slightly rounded to ensure that the curve is differentiable. By making the middle section steeper and steeper, the time when the velocity is non-zero can be made as small as we please. But again there is no curve for which the velocity is always zero, and so the problem has no solution.

the impact his raising the problem had on the study of the calculus of variations. Did mathematicians respond to his challenge by taking up this, as he saw it, somewhat neglected field, and if they did, did they do so in ways that reflected his ideas? A partial affirmative answer is provided by Hilbert himself, who in a paper (1906) took up and extended the ideas given in his account of the 23rd problem. A response of a different kind was given by the German mathematician W. Ritz in 1909, who developed ideas given earlier by the English physicist Lord Rayleigh (and unknown to both of

Box 4.7 A 'proof' that 2 = $\sqrt{2}$

Consider the right-angled triangle AOB in the figure below. The lengths *AO* and *OB* are each 1, so the length of *AOB* is 2, and the length of the hypotenuse *AB* is $\sqrt{2}$. Now consider the broken curve $AO_1B_1O_2B$. Clearly its total length is also 2, and so is the length of the zig-zag curve $AO_3A_1O_4B_1O_5A_2B_2B$. Carry on forming zig-zag curves in this fashion.

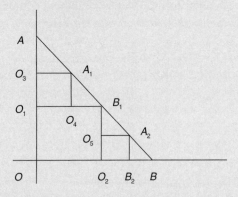

The zig-zag curves get as close as you like to the hypotenuse. If it followed that the length of the zig-zag curves got arbitrarily close to the length of the hypotenuse, it would follow that the limit of a sequence of 2s was $\sqrt{2}$. That is absurd, but the result is that there has to be a sound mathematical way of saying that no zig-zag curve, however close to the hypotenuse, provides a good approximation to the length of the hypotenuse.

them in the 1860s by the French mathematician Liouville).[51] What became known as the Rayleigh–Ritz method was a method of successive approximations particularly well suited to finding extremals, but the theoretical task of showing that approximations actually converge to a minimum turned out to be very difficult.

---

[51] J. Lützen, (1990).

The theoretical task was taken up by mathematicians who learned the subject from Kneser's book. Consider the simplest case of a variational problem, in which the solution is a curve in the plane. For example, let the problem be to find the path of light in a certain medium from one given point to another (the target). The first problem is to find the path of the 'shortest' ray of light from the source to the target. The second problem allows the target to move along a curve, and asks for the path of the ray of light in this more general setting.

Kneser's approach to this problem was deliberately modelled on the theory of geodesics which Gauss had published in the 1820s. Gauss showed that one can draw something very analogous to polar coordinates on any surface (Figure 5). First, pick a point, O and draw an arbitrary geodesic through it. All other geodesics make a specific angle with the first one, so one can attach the 'geodesic polar coordinates' (r,θ) to the point P by saying it is the point that lies a distance r along the geodesic making an angle θ with the first geodesic. This gives unique coordinates in any region where the geodesics only meet at the point O, and every point in such a region of the surface acquires coordinates in this fashion. For example, if on the Earth the North Pole is taken as the point O then every point except the South Pole acquires unique coordinates. The geodesics through O are the lines of longitude. Gauss showed that on any surface the curve composed of all points with coordinates (r,θ) for a fixed value of r (the 'circle' centre O of radius r, if you

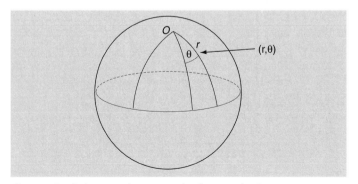

Figure 5 Geodesics on a sphere emanating from a point O.

like) meets the geodesics through O at right angles. In the case of the Earth, these curves are the circles of latitude, including the equator.

Another method that Gauss described is to draw a geodesic γ through an arbitrary point O and to consider all the geodesics perpendicular to γ (Figure 6). One can attach the 'geodesic co-ordinates' (u,v) to the point P by saying it is the point that lies a distance v along the geodesic perpendicular to the geodesic γ and crossing it a point a distance u from O.

This method also works for pieces of the surface that are not too large. Any curve composed of all points with coordinates (u,v) for a fixed but arbitrary v (the curves everywhere a distance v from the original geodesic γ) meets the geodesics emanating from γ at right angles. On the Earth, we could choose the equator as the geodesic γ, the geodesics coming from it at right angles would be the longitudes, and the equidistant curves would be the circles of latitude. In either case we have a one-parameter family of geodesics and a one-parameter family of curves (which are not geodesics) meeting the geodesics at right angles. These curves are called transversals.

Now it is of course true that on a surface the equation of a geodesic is obtained from a variational principle. Kneser's idea was to see how far you could get with any variational problem of a suitably similar form. He was able to recapture much of Gauss's work in this more general setting, to solve variational problems where the end-point is also allowed to vary, and to show that his extremals are genuine minima.

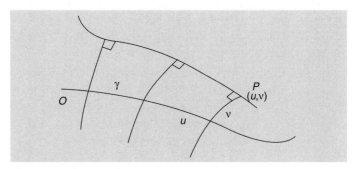

Figure 6 Geodesic coordinates.

Box 4.8 Kneser on the calculus of variations

Kneser's idea treated any variational problem of the form $J = \int F(x,y,x',y')dt$, where x and y are functions of t, as if it were a question about geodesics. Any candidate $x(t)$, $y(t)$ can be thought of as a curve in a part of the plane. Kneser began by taking a one-parameter family of extremals for J. These are curves $(x(t,a),y(t,a))$ for varying t which, for each fixed value of the parameter a in some interval, are extremals. He showed that for a wide range of problems the curves $(x(t,a),y(t,a))$ for varying a and fixed t form a family of transversals. In this case each point of the region of the plane where the extremals and their transversals are defined can be given new coordinates (t,a).

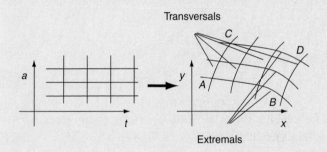

Transversals

Extremals

He then showed that the integral J taken along two extremals in the family, say AB and CD between the same transversals AC and BD, has the same value. This was still true if one of the transversals was a point, in which case the extremals emanate from that point (analogous to the first of Gauss's cases).

He then found conditions under which the parameters t and a can be used as coordinates, the t-curves being the extremals and the a-curves the transversals. From this he deduced that if A and B are two points on an extremal, then the integral J taken along any rival curve joining the point B to any point on the transversal is minimised by the extremal C_0.

This solves a variational problem where the end-point is also allowed to vary. Second, he could show that his extremals are genuine minima.

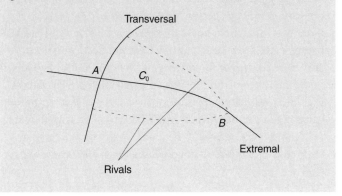

Hilbert's direct method, as it became known, proceeds by analogy with the questions:

1. find the minimum value of a given function on the interval $(0, 1) = \{x : 0 < x < 1\}$, and

2. find the minimum value of a given function on the interval $[0, 1] = \{x : 0 \leqslant x \leqslant 1\}$.

In the first case, if we are given the function $f(x) = 2x$ there will be no minimum value, even though the function is continuous and bounded. The function takes values arbitrarily close to 0, but never the value 0 itself. But in the second case any continuous function is not only bounded but it attains its bounds, and in particular its minimum values. The difference between the two problems is that at least for well-behaved functions, the second interval is somehow better.

What Hilbert proposed was to consider variational problems in the same spirit. An integral $J = \int F(x,y,x',y')\,dt$ is given, involving a specified function F, and the candidates for extremals are curves $(x(t,a),y(t,a))$. What could go wrong? It could be that as a curve $(x(t,a),y(t,a))$ is altered into one for which the corresponding value of J is less, the new curve ceases to have a tangent everywhere, and perhaps in the limit it has no tangent anywhere. Hilbert showed that, under certain conditions of the function F, this posed no problem: the integral J still made sense and the limiting curve not only existed but was one where the integral J attained a minimal value.

So if it was required to find a curve between two points for which the integral J took a minimum, Hilbert proposed to consider all curves of finite length joining the given points and to assert, if certain conditions of the function F were fulfilled, that there was definitely a curve along which J attained its minimum. He could also show that, under certain other conditions, the minimising curve was typically quite smooth, even though it was necessary to consider highly non-differentiable curves in the course of the proof.

The subsequent history of the calculus of variations reveals a virtual explosion of interest in this area. Jesse Douglas's brilliant exploitation of the direct method was enough to win him one of the first Fields Medals in 1936 (the Fields Medals are generally regarded as the equivalent of a Nobel Prize in mathematics). The subject has merged with questions in the theory of partial differential equations, functional analysis, and other branches of mathematics, and it would be impossible to survey it in less than a book. How much of the credit for this is due to Hilbert, and in particular to his choice of Problems, is therefore impossible to say, but surely it was considerable. His direct method became, and remains, one of the most powerful approaches, and mathematicians were won to the study of the subject. His express intention was to bring to the subject the attention it was due, and in that he succeeded.

Problems 21 and 22

The questions Hilbert asked about the Riemann–Hilbert problem (number 21) and the uniformisation theorem (number 22) were speedily resolved. It is best to take them in reverse order (because the 22nd is in some ways a special case of the 21st). In raising the question of uniformisation at all, Hilbert was not criticising Poincaré, but drawing attention to the truly fundamental nature of the result. The uniformisation theorem, if it could be established, was bound to be a result of major significance in complex analysis, and complex analysis had established itself during the nineteenth century as one of the major parts of mathematics.

The subject of complex analysis is very different from real analysis. Perhaps the most striking difference, and the one that presumably led to Hilbert placing it where he did in his list of problems, is that complex analysis can be seen as the paradigm case of the phenomena he had pointed to in Problem 19, because complex analytic functions are the paradigm examples of the unexpectedly smooth functions that satisfy a simple elliptic system of partial differential equation (the so-called Cauchy–Riemann equations). These are the equations satisfied by the real and imaginary parts u and v respectively of a function $f(x,y) = u(x,y) + iv(x,y)$:

$$\frac{\partial u}{\partial x} = \frac{\partial v}{\partial y}, \quad \frac{\partial v}{\partial x} = -\frac{\partial u}{\partial y}$$

It follows that the u and v are what are called harmonic functions: they satisfy the partial differential equation

$$\frac{\partial^2 z}{\partial x^2} + \frac{\partial^2 z}{\partial y^2} = 0.$$

The remarkable feature of an analytic function is that once its values are known on a patch, however small, they are known on the largest region where that function can be defined and which contains the patch.[52]

Poincaré's remarkable uniformisation theorem is the claim that any algebraic curve can be uniformised, provided suitable (new) functions are introduced.[53] This was already known for non-singular cubic curves, but for the general curve the result was totally unexpected. (Poincaré's claim in 1883 that any expression of the form $f(z,w) = 0$ can be uniformised was even more striking.)

The uniformisation theorem says even more. It says that for algebraic curves the functions which arise are either defined on the

---

[52] This is the property of analytic continuation that underlies Hilbert's discussion of Problem 18 and 'that pliant instrument, the power series'.

[53] The new functions include the Fuchsian ç-functions which Hilbert mentioned briefly.

Box 4.9 Uniformisation

Each point of the circle, $x^2 + y^2 = 1$ can be represented parametrically as $(\cos t,\ \sin t)$. The circle is said (somewhat awkwardly) to have been uniformised by the trigonometric functions cos and sin, because the coordinates of each point of the curve are expressed as single-valued functions of a variable.

There is another uniformisation, which is in some ways more striking. The straight line through the point $(0, -1)$ with equation $y = mx - 1$ meets the circle again at the point $(s, ms - 1)$ and since $s^2 + (ms - 1)^2 = 1$ it follows that either $s = 0$ or

$$s = \frac{2m}{(1 + m^2)}.$$

So points on the circle have the form

$$\left(\frac{2m}{1 + m^2},\ \frac{m^2 - 1}{m^2 + 1} \right),$$

which is also familiar from elementary coordinate geometry. In this case each coordinate is a rational function of the parameter m, and the curve is said to have been uniformised by rational functions.

Uniformisation also works in the complex case, where equations give rise to Riemann surfaces. The equation $z^2 + w^2 = 1$, for example, has a Riemann surface with two points for every z except $z = \pm 1$, where $w = 0$. An analysis of what happens near the points $z = \pm 1$ shows that the values of w are interchanged as z goes round either of the points. The uniformisation

$$\left(\frac{2m}{1 + m^2},\ \frac{m^2 - 1}{m^2 + 1} \right),$$

where m is now thought of as a complex number, expresses the points in the Riemann surface of $z^2 + w^2 = 1$ as single-valued functions of m.

Riemann sphere (and are rational functions) or they are defined on the entire complex plane (and are elliptic functions) or are defined on the unit disc, in which case they belong to the new class of functions that Poincaré defined in a series of long, and at times obscure, papers in the 1880s, right at the start of his career. So there is an intimate connection between Euclidean geometry, spherical geometry, and non-Euclidean geometry on the one hand and the theory of complex functions on the other.

Poincaré's original argument in 1883 had proved one half of the uniformisation theorem. He successfully showed how the geometry comes in, and made it plausible that the coordinates of the points on the original surface will be single-valued functions of a variable free to roam over the appropriate domain. But his proof that the parameterisation was analytic, as the uniformisation theorem claimed, was flawed, as Hilbert noted. It broke down at infinitely many points on the boundary of the disc. Poincaré admitted this when he took up the question again in 1907, and give two ways of dealing with it. 'The problem,' he said, 'is none other than the Dirichlet problem applied to a Riemann surface with infinitely many leaves.' He now reworked his earlier approach, making heavy use of the techniques of potential theory, a subject of which he was a master, and dealt explicitly with the problem Hilbert had highlighted.

His paper came out at the same time that Paul Koebe, a doctoral student of Schwarz's in Berlin, began his work. Koebe published prolifically, pursuing a ruthless strategy to gain the attention of the leading mathematicians in Germany. He solved the problem of uniformising algebraic and analytic curves in general and in many particular cases where there are special features of the problem that require attention. He also set about making the old 'continuity method' of Klein and Poincaré rigorous. He made a point of contrasting the merits of his work with Poincaré's approach, showing how this or that technical result could be avoided. When Hilbert revived Dirichlet's principle Koebe responded with an explanation of how these ideas lay close to his own. He published at length in the *Göttingen Nachrichten*, making sure his papers were presented by Hilbert and Klein, in *Mathematische Annalen*, and in the Berlin-based *Journal für Mathematik*, and the French *Comptes Rendus*. Not surprisingly, he was invited to speak to the ICM in Rome in 1908 on the solution of Hilbert's 21st problem, quite an honour for a young man.

The importance of his work could not be doubted, although the usual response was attempts to give shorter proofs of what Koebe had presented in this torrent of lengthy papers. His work was made the focus of a special meeting of the DMV in 1911.[54] Brouwer described how his recent proof of the invariance of dimension fitted into the problem, and enabled one to rescue the dubious continuity method introduced by Klein and Poincaré, and illustrated it with reference to the Fuchsian case. Koebe replied that he had been able to extend this to other cases, and gave an example of a special case that Riemann had himself considered where, he said, Poincaré's methods could not work. However, there was in fact quite a fierce disagreement between Koebe and Brouwer about the extent to which Koebe's work on the continuity method was valid.[55] Koebe thought it was completely valid, Brouwer that it was restricted to just the cases covered less precisely by Poincaré. A subsequent exchange of letters between these two forceful personalities revealed that Koebe repeated his claim but could not come up with a detailed proof, and it was a long time before he was willing to accept the justice of Brouwer's position. Others too did not like to commit themselves, and it was not until 1923 that Klein acknowledged the essential contribution that Brouwer made to the vindication of the continuity method he and Poincaré had introduced some 40 years before.

With this complete solution, the topic fell asleep until just before the Second World War, when the German mathematician Oswald Teichmüller started to re-write it in a way that allowed mathematicians to study the moduli or parameters upon which an algebraic curve depends. Teichmüller, a fanatical member of the Nazi Party, disappeared on the eastern Front in 1943, but a number of mathematicians in the group around Lars Ahlfors at Harvard and

---

[54] See the *JDMV*, **21**, 153–166.

[55] See van Dalen (1999), pp. 180–193.

Lipman Bers in New York took it up after the War. Bers finessed the obvious problem posed by associating even indirectly with Teichmüller by quoting from Plutarch's *Lives* 'it does not of necessity follow that, if the work delights you with its grace, the one who wrought it is worthy of your esteem.' One of their achievements was to lead mathematicians back to the original ideas of Poincaré. He had produced a theory of the groups that act on 2- and 3-dimensional non-Euclidean space and produce Riemann surfaces. The theory in the 3–dimensional case lay more or less where Poincaré had left it, and where Klein and his indefatigable assistant Fricke had left it before the First World War, in a morass of partly proved claims served up in a way that seemed to defy any attempt to make them rigorous. The two-volume work by Fricke and Klein, for example, was known in the 1960s as the Book of Revelations because of its obscurity. Teichmüller theory, in the hands of Ahlfors and Bers, has offered a new approach that supplies a good deal of the much-needed clarity.

Hilbert had also called for the uniformisation theorem to be extended to functions of three or more variables, and he pointed out that several particular cases were already known. But if he imagined that these cases were typical, and precisely analogous results would be found, then he was unlucky. There is no uniformisation theorem for three or more variables.

The Riemann–Hilbert problem is a cousin of the uniformisation theorem. It concerns complex functions which are defined by differential equations. In 1857 Riemann had shown how to obtain a group of matrices from a specific linear differential equation. He then asked himself if, conversely, the matrices determined the differential equation, and he showed in a posthumously published work that they did.[56] Lazarus Fuchs showed in Berlin in 1866 that there was a recognisable class of differential equations that generalised the one Riemann had studied, and they too gave rise to a group of matrices. He then asked the converse question: do the matrices determine the differential equation? But neither he nor

---

[56] It is this group of matrices that form the monodromic group of the equation, to use the term introduced by Hermite and used by Hilbert. See Gray (1999).

any one else could solve the problem. In the early 1880s Poincaré took up Fuchs's work, but his approach was so bold (or so vague, to offer another opinion of the day) that it was not regarded as satisfactory. Towards the end of the century Fuchs's student Ludwig Schlesinger then took up the problem and attempted to make Poincaré's method rigorous.

In 1904, in a short paper presented to the ICM at Heidelberg, Hilbert showed how a related question could be tackled using Fredholm's newly discovered method of integral equations. This insight was then taken up as the basis of Plemelj's more thorough treatment of the Riemann–Hilbert problem, as Hilbert's 21st Problem was known.[57] Plemelj formulated the Riemann–Hilbert problem as a question about an integral equation, using the technique pioneered by Fredholm, and his affirmative answer was to be regarded as definitive for 85 years.

Five years later Plemelj's methods received independent confirmation when the young American mathematician G.D. Birkhoff published an article in which he solved the Riemann–Hilbert problem almost without realising it.[58] Only at the end of the journal did Birkhoff add a note to say that he had now noticed, as Toeplitz had written to him to point out, that his proof not only amounted to a proof of the Riemann–Hilbert problem, previously solved by Hilbert and Plemelj, but that his was indeed 'not markedly different from theirs.' He returned to the problem later that year, and simplified the treatment of Hilbert and its 'elegant completion' by Plemelj by eliminating the use of Fredholm theory and spelling out the details of the connection with the Riemann–Hilbert problem. At this point every one agreed that Hilbert's 21st Problem was completely solved, and so matters were to remain until 1989.

In the 1980s, Russian mathematicians working in the theory of differential equations became suspicious, but it was still a shock when Anosov and Bolibruch announced in 1989 that the problem

---

[57] Josef Plemelj (1873–1967) studied in Berlin and Göttingen before teaching at the University of Vienna. After the First World war he returned as a Professor to his native Yugoslavia.

[58] Birkhoff (1913).

had not been solved, and the Riemann–Hilbert problem is to be answered correctly in the negative. They showed, both on theoretical grounds and with a whole family of counter-examples, that it is possible to specify singular points and matrices in such a way that there is no corresponding differential equation.[59]

How could it be that a mistake of this kind could survive for so long? Part of the answer is that it was, like the purloined letter, hidden where no-one would look for it, in a body of published work every one 'knew' was completely rigorous. But another part of the answer is that Hilbert's formulation of the Problem, when he raised it, in Paris, was vague, and as Anosov and Bolibruch pointed out, Hilbert's remark needs careful interpretation if it is to make sense. Hilbert had posed the question, for good reasons, as a question about a system of differential equations. There is, moreover, a good way of passing from a differential equation of order 3, say, to 3 first-order differential equations. And Fuchs' criteria for a differential equation to be one for which the Riemann–Hilbert problem makes sense can be formulated for systems too. But the criteria are not the same: there are differential equations of Fuchs' type that do not give rise to a Fuchsian system of differential equations. This subtle distinction led Anosov and Bolibruch to their counter-examples: a group of 3×3 of matrices with no corresponding 3rd-order differential equation.

Hilbert Problems on number theory

One measure of the success of a group of people is that it provides for its own succession. Hilbert did this in number theory, but almost by accident. He supervised a number of the leading number theorists of the next generation, although it was in a somewhat distant way, and eventually Göttingen was able to appoint Helmut Hasse as its leading specialist in the field. But Hasse was scarcely a product of Göttingen, and the techniques he used were derived

---

[59] The matter is presented most accessibly in their book, Anosov and Bolibruch (1994), which briefly indicates how the gap in Plemelj's proof was noticed and exploited by a number of Soviet mathematicians (see pp. 8–9).

from mathematicians more closely tied to Berlin and the work of Kronecker. Similarly, Takagi, the first Japanese mathematician of truly international standing, was only loosely connected to Göttingen. But their successes with Hilbert's Problems cemented their reputations and ensured the continued growth of algebraic number theory in the twentieth century.

Problems 9, 11, and 12—The comedy of errors

Minkowski's complicated theory of quadratic forms, the background to the 11th Problem, was not supplanted until Helmut Hasse propounded a new theory in 1923–24. Hasse had studied under Hilbert's pupil Erich Hecke briefly in Göttingen before going to Marburg to study under Hensel. This was not an obvious career move, and Hasse had only become interested in Hensel's approach to number theory on browsing through Hensel's book in a Göttingen book shop. But it was the right move, for Hasse saw that Hensel's ideas were likely to answer many deep questions in the subject.

Hensel's theory is technical, but it allows a mathematician to mimic the way the rational numbers can be completed to form the real numbers. The usual way that is done allows mathematicians to say that every sequence of numbers that looks as if it is converging actually does converge to a number. It allows mathematicians to say what π is, when all we can write down is successive approximations to it, such as 3, 3.1, 3.14, . . . Hensel found other ways of completing the rational numbers, one for each prime number p. This allowed mathematicians to reap some of the benefits of modular arithmetic and some of the benefits of analysis.

Hasse showed that two quadratic forms over the rationals are equivalent if and only if they are equivalent over each completion, including ∞. This looks as though infinitely many things have to be checked, and all of them involve new and unfamiliar mathematics. However, it is actually a great breakthrough. For the usual completion, which replaces the rational numbers by the real numbers, an old theorem due to Sylvester does the trick. For the others Hasse showed how to use the Hilbert symbol. In fact, the problem reduces to finitely many cases, all of which are easy. In the jargon of the trade, Hasse showed that a global question was equivalent to infinitely many local questions.[60]

I said in Chapter 3 that Hilbert's claim in Problem 12 was wrong, and that our question would be to see whether the influence of Hilbert contributed to the confusion. Hilbert himself did very little directly on this Problem. But he had a Swiss student called Rudolf Fueter who did, and his former Professor Heinrich Weber in Strasburg had another, Daniel Bauer. Fueter, Weber and Bauer worked hard on the problem, which rapidly became very technical, and Fueter claimed to have shown in two papers of 1905 and 1907 that Hilbert's claim about how certain fields are generated is true. This claim was even endorsed by Erich Hecke in the preface to his thesis, written under Hilbert in 1912, although he did add a footnote to say that Fueter will fill a few gaps in his papers when he publishes his book on the subject.[61]

This consensus lasted until 1914, when Fueter realised he had a counter-example to Hilbert's claim. He could then guess what the correct answer would be. But he fumbled his opportunity, and after the War it was a Japanese mathematician, Teiji Takagi, who put the new picture together in all its detail in 1920. Takagi may have started serious work on the problem in 1914, when, as he said, he expected the flow of books and journals from Germany to dry up. But he was not a complete outsider. He had studied in Berlin in 1898, and in Göttingen in the Spring of 1900, and Hilbert had supervised his thesis, which was successfully submitted to the University of Tokyo in 1903. He later recalled that Hilbert had been not very directly involved with his thesis work, but he took care in his thesis to spell out the close connections it had with the 12th Problem. It would be hard to say whether it was the difference between Fueter and Takagi as mathematicians that made the difference, or if Takagi's greater distance from Hilbert enabled him to

---

[60] A more delicate question is to investigate quadratic forms allowing only changes of variables that have integer coefficients. In the mid-1930s Siegel showed that Hasse's local–global principle breaks down, and established a more complicated and subtle theory.

[61] This account is based on Schappacher (1998).

Box 4.10 How Hilbert's symbol generalises the Legendre symbol

Hilbert's norm residue symbol is defined for an arbitrary number field K to be

$$\left(\frac{a,b}{P}\right) = \begin{cases} 1, & \text{if } ax^2 + by^2 = 1 \text{ has a solution } x,y \in K_p, \\ -1, & \text{if not.} \end{cases}$$

Here a and b are non-zero elements of K, and K_p denotes what is called the completion of K at P. We shall not define it, but the point to hang on to is that K_p is an easy field to work in. The key result about the norm residue symbol is

$$\prod_P \left(\frac{a,b}{P}\right) = 1.$$

Let $K = \mathbf{Q}$, the rational field. Then for all primes r other than 2, p, q, and ∞ the symbol $\left(\dfrac{p,q}{r}\right)$ must take the value 1. So

$$\prod_P \left(\frac{a,b}{P}\right) = \left(\frac{p,q}{\infty}\right)\left(\frac{p,q}{2}\right)\left(\frac{p,q}{p}\right)\left(\frac{p,q}{q}\right) = 1.$$

But $\left(\dfrac{p,q}{\infty}\right) = 1$, because $\mathbf{Q}_\infty = \mathbf{R}$ and p and q are positive. So

$$\left(\frac{p,q}{p}\right)\left(\frac{p,q}{q}\right) = \left(\frac{p,q}{2}\right).$$

A simple but technical argument which we omit shows that

$$\left(\frac{p,q}{2}\right) = (-1)^{(p-1/2)(q-1/2)},$$

the value of the Legendre symbol in this case.

grasp the prize. But the impression must be that on this occasion Hilbert's influence delayed the solution of the 12th Problem.

Takagi's work led, almost by accident, to a solution of Hilbert's 9th Problem. In his 9th problem, Hilbert had called for a reciprocity theorem for arbitrary powers in any number field, extending work he had begun in his *Report*. In the 1920s Takagi developed a general theory of number fields which are a particular kind of field extension with Abelian Galois group. For technical reasons the restriction to Abelian extensions is quite fundamental.[62] This work did not, however, bring with it a reciprocity law. But in 1923 Emil Artin conjectured a result which, while seeming quite abstract, yielded all known reciprocity laws when it was made explicit. He did not arrive at it by studying Takagi's work, nor was he trying to solve a Hilbert problem. Instead, he was doing analysis, and studying some generalised L-series which he had defined. However, using Takagi's theory and some ideas due to the Russian mathematician Chebotaryev, Artin was able to establish his conjectures in 1927, and his work soon led Hasse to define a generalised norm residue symbol and thus complete the solution of Hilbert's 9th problem.

All this work can be said to go back to Hilbert's pioneering studies of quadratic fields, which are Abelian field extensions, and all those who worked in the area acknowledged his influence. By the time the relevant Hilbert Problems were solved, voices could be heard saying that his famous *Report* was a rather one-sided presentation, and indeed it was, but it set the research agenda for a generation, and must be accounted one of Hilbert's greater successes.[63]

Hilbert's geometric Problems

As we shall see, these questions proved to be highly uneven. Some, like the 13th , 14th and 17th, fell quiet. Others, like the 15th, led to a prodigious amount of mathematics, but probably not because of any contribution of Hilbert's. Two did enjoy some success in the period 1900–1918, the first of the two that formed the 16th Problem, and one of the three that formed the 18th. Even then, the prestige that accrued was mixed, and while the woman who deepened our understanding of the 16th Problem retreated into obscurity the man who solved the 18th embarked on a career that led to permanent notoriety.

---

[62] A conjecture by Jean-Pierre Serre about non-Abelian extensions has been called the Jugendtraum of the twentieth century.

[63] See Zassenhaus's discussion in Rüdenberg and Zassenhaus (1973), pp. 17–21.

Ragsdale and others on Hilbert's 16th Problem

Part of Hilbert's 16th Problem asked about the number of loops a real algebraic curve of a given degree can have, and how they can be arranged. This pointed the way to the study of real algebraic varieties, for which the known methods of algebraic geometry were inadequate, but even so Hilbert may not have expected that it would be as long as it has been for a general theory to emerge appropriate to the kind of problems he had in mind. Indeed, it can be argued that such a theory is only being created at the end of the century.

Two of Hilbert's students Grete Kahn and Klara Löbenstein, wrote PhDs on the subject, both in 1909. So too did a number of Americans, J.E. Wright and Virginia Ragsdale among them. Much of this work focused on the obscure and unproved claim made by Hilbert about curves of degree 6, and for some time the eventual winner in this competition was adjudged to be Karl Rohn's contribution of 1911, although this was eventually shown to be unsatisfactory by the Russian mathematician D.A. Gudkov in 1954. However, Ragsdale's work was more general and merits some discussion.

Virginia Ragsdale was born in Jamestown Virginia on 13 December 1870. In 1892 she won a scholarship to Bryn Mawr University, and after a year of graduate study there the opportunity of a foreign fellowship enabled her to take Charlotte Scott's advice and study for a year at Göttingen under Klein and Hilbert. She returned to teach for several more years at Bryn Mawr School in Baltimore before a scholarship took her to back to Bryn Mawr itself, where she built upon her work in Germany to complete her PhD in 1906. It was published in 1906, and seems to have been her only publication. She eventually became a professor at the Woman's College in North Carolina, where she remained from 1911 to 1928 and was regarded as an excellent, patient teacher with high standards. She died on 4 June 1945.[64]

---

[64] I am indebted to the archivist of Bryn Mawr for sending me the information on which this account of Ragsdale's life is based, some of which was published by the Women's College of North Carolina in 1946.

Her thesis, which one must presume was written under the supervision of Charlotte Scott, is entitled 'On the arrangement of the Real Branches of Plane Algebraic Curves'. In it she began by surveying what was already known, observing that Hilbert himself had never published any details of the 'extraordinarily circumstantial' process by which he had come to some of his conclusions. She found that two methods were known for generating curves of even order with the maximal number of real branches: a simple one due to Harnack's, which gave one arrangement, and a more complicated one due to Hilbert, which gave a more complicated arrangement. She then introduced a modified form of Hilbert's method. By a careful analysis of many cases, she was led to claim that if a non-singular curve of degree $2n$ has the maximum number of real branches, then at least $(1/2)(n-1)(n-2)$ of its $(2n-1)(n-1)+1$ ovals must be internal, and not more than $n^2 + (1/2)(n-1)(n-2)$ of them can be external. This generalises Hilbert's unproved claim that if a non-singular curve of degree 6 has the maximum number of branches then at least one of its 11 ovals must be internal. This claim was then proved by J.E. Wright in 1906.[65]

All the methods that were used were based on generalising the idea that a hyperbola can be obtained by slightly varying the co-efficients in the equation that defines the degenerate curve composed of two lines. For example, from the second degree equation $xy = 0$ which defines the lines with two equations $x = 0$ and $y = 0$ each of the first degree, one obtains the equation $xy = a$ which defines a hyperbola. Similarly, a general non-singular curve should be obtained by slightly varying the coefficients in a degenerate curve composed of a curve of degree $m - k$ and one of degree k. The curve of degree $m - k$ is chosen to have the maximum number of circuits for one of that degree, and the auxiliary curve, of degree $m - k$, is to be chosen astutely, so that the new curve has as many more loops as possible. Harnack's method was based on the systematic study of a curve of odd degree and a line, Hilbert's on a curve of odd degree and an ellipse; Ragsdale took up the case

---

[65] Wright (1906).

of odd degree and a line pair, which was intermediate between the two.

Rather than follow her detailed account, we content ourselves with two sets of figures taken from her paper, see figure 10. The first shows the sort of arrangements that can occur for curves of low degree, and it is striking that many possibilities are excluded. The second, which she stressed was only qualitatively accurate, gives a hint of the method of variation of coefficients as used first by Harnack and second by Hilbert.

In a concluding passage she showed how her ideas could be reformulated using Dyck's theory of the characteristic number associated with a manifold, and observed that there may therefore be some underlying relation to the theory of multiply-connected surfaces. This shows how closely she was wedded to her largely algebraic theory of real curves, for the view from Göttingen was that the natural way to approach these questions was via the genus of the underlying Riemann surface. Indeed, in his paper of 1876 which started the whole subject off, Harnack stated the purpose of his paper in these terms: 'A curve of genus p never has more than $p + 1$ separated closed paths and indeed that it is also true that for every value of the genus p there are curves with $p + 1$ loops.'

In 1939 the Russian mathematician Petrovskii proved one of Ragsdale's weaker conjectures, seemingly without knowing of her paper, and proposed some conjectures of his own.[66] He established Hilbert's claim that there is no curve of degree 6 all of whose ovals lie outside each other, and showed more generally that an algebraic curve of even degree $2n$ has at most $3/2(3n^2 - n) + 1$ ovals lying outside each other. In 1969 Gudkov came up with a new arrangement of ovals that Hilbert had missed, and also formulated some new conjectures. Then in 1979 the Russian mathematician Oleg Viro proved that Ragsdale's conjectures were in fact false, and in 1993 his student Ilia Itenberg showed they were in fact quite significantly misleading.

In their joint account Itenberg and Viro (1996) were full of

---

[66] This account follows Olejnik's in Alexandrov (1979), pp. 233–249.

praise for Ragsdale's pioneering achievement. She had directed attention to curves of even degree, introduced the distinction between odd and even ovals (an oval is even if it lies inside an even number of other ovals, otherwise odd) and focused attention on the number, p, of even ovals and the number, n, of odd ovals. Her strongest conjectures about these numbers were that for a curve of even degree $2k$

$$p \leqslant \frac{3k(k-1)}{2} + 1 \text{ and } n \leqslant \frac{3k(k-1)}{2} ,$$

while more cautiously she also conjectured that

$$p - n \leqslant \frac{3k(k-1)}{2} + 1 \text{ and } n - p \leqslant \frac{3k(k-1)}{2} .$$

The first of these more cautious conjectures was also established by Petrovskii in 1939, with a proof that follows the lines of her argument in support of the conjecture, although it is clear he did not know of her work.

There is no space here to describe the details of Itenberg and Viro's work but a picture may suffice.[67] They begin with a large square divided into triangles. Each vertex of the triangulation comes with a plus or minus sign. They then give a rule which draws a line in some of these triangles from one mid-edge point to another, depending on the signs of the vertices. The hard work is in then showing that the diagram so obtained is a picture of a real curve defined by a polynomial equation with real coefficients. That being achieved, astute choice of signs permitted them to construct counter-examples even to the strongest of Ragsdale's conjectures. In fact, their pictures are qualitatively accurate pictures of the corresponding real curve.

Figure 7 shows what the simplest curve of Harnack's type looks like in this description. Figures 8 and 9 show counter-examples to Ragsdale's conjectures for p and for n, and then for the arrangements of the corresponding ovals (Figure 10).

---

[67] The paper by Itenberg and Viro (1996) is very clear.

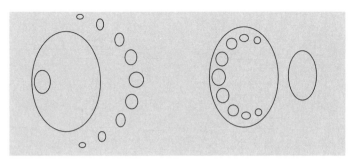

Figure 7

Figure 8

Their main result (due to Itenberg) is that there is a non-singular curve of even degree $2k$ for which

$$p = \frac{3k(k-1)}{2} + 1 + \left[\frac{(k-3)^2 + 4}{8} \right]$$

Figure 9

Figure 10

and a non-singular curve of even degree $2k$ for which

$$n = \frac{3k(k-1)}{2} + 1 + \left[\frac{(k-3)^2 + 4}{8} \right],$$

where in each case the square brackets denote 'the integer part of' the expression contained by them, so it only becomes positive when $k \geqslant 5$. So there are counter-examples to Ragsdale's conjectures for curves of degree 10 or more, but not less.

They also point out that two of Ragsdale's conjectures still stand. These assert that for a curve with the maximum number of ovals

$$p \geqslant \frac{(k-1)(k-2)}{2} \text{ and } n \geqslant \frac{(k-1)(k-2)}{2}.$$

Bieberbach on Hilbert's 18th Problem

Hilbert directed attention to three questions in particular. First, he asked if it is the case that Euclidean n-dimensional space admits only finitely many patterns? This was soon answered in the affirmative by the opportunistic and power-seeking Bieberbach in 1910, who also gave estimates of this number in low-dimensional cases that confirmed the known results: there are just 17 patterns of this kind for the Euclidean plane, and 219 patterns, or crystal structures, for Euclidean 3-dimensional space. Never one to miss an opportunity to ingratiate himself, he published his solution in a Göttingen journal although he was officially based in Berlin.[68]

The second of the 18th Problems, so to speak, asked if there are regions that can tile a plane but are not the fundamental domains of a group. Remarkably enough, the answer is that there are. The first example was actually 3–dimensional: Reinhardt described a polyhedron that fits together to fill 3–space but does not form a fundamental domain. Then in 1932 the topologist Heesch, who was then at Göttingen, found an example in the plane.[69] It is shown in Figure 11. The trick is to construct a tile that can be flipped over and fitted alongside its earlier position. The two tiles form a

---

[68] Bieberbach (1910).
[69] See Möhring (1998).

Figure 11

fundamental domain for the group of symmetries of the pattern, but the tile itself, Heesch showed, is not itself a fundamental domain. It is pleasant to record that the German ceramics firm Villeroy and Boch then made samples of the tile to Heesch's specifications (he produced a figure with 12 vertices) which were then incorporated into the roof of the Göttingen Stadhaus, which now houses the town library (see Figure 12).

Hilbert had also asked about what can be done when the regions do not fit together completely but leave gaps. How well can spheres of unit radius be packed together, for example, in the sense that the gaps are as small as possible? Everyone expects that the cannonball arrangement will prove to be best possible in 3-dimensional space,

Figure 12

but even in 1999 this has still not been proved (although a final proof is thought to be close).[70] In his report on the problem in 1974 Milnor called the situation scandalous. In spaces of higher dimensions the problem might seem to be merely arcane, but it has a natural interpretation in digital communication theory. A sample of a signal can be thought of a list of numbers, and therefore a vector in some high-dimensional space. Nearly equal signals appear as nearby vectors, so to keep them distinct Shannon, the father of this entire theory, suggested imagining each signal enclosed in a sphere, whence the sphere-packing problem entered the modern world. Moreover, 24-dimensional space generated one of those unexpected twists that makes mathematics surprising even to the experts. There is a way of packing spheres together in this space that comes very close to what can be shown to be the best possible way. That means that there is a highly symmetric lattice in this dimension, whose points lie at the centres of the

---

[70] Sloane (1998).

spheres. This lattice, called the Leech lattice, has so far played an important role in the classification theorem for finite groups, and in the elucidation of a connection between the representation theory of these groups and the theory of modular forms. Modular forms are a topic with roots in the Jugendtraum and Hilbert's 12th problem; the connection would surely have delighted Hilbert.

CHAPTER 5

Between the wars: foundations examined

Göttingen after the First World First World War; the eclipse expedition

1918–1933

After the First World War, the shattered nations of Europe tried to rebuild. Germany was forced to pay massive compensation to the victorious Allies, who blamed it for starting the war and causing so many deaths. This deepened the collective sense of failure and defeat brought to a nation that, as late as Spring 1918, had thought it could still win the war. Turmoil to the east, in the newly formed and revolutionary Soviet Union, spilled over into violent social unrest from Right and Left in Germany, and when that stabilised the Weimar Republic had to ride out several years of hyper-inflation. Germany's relations with its neighbours were poisoned by the war for many years. Many international organisations, especially those with a strong French component, sought to exclude or marginalise Germany, and the country was not admitted to the League of Nations. The first International Congress of Mathematicians to be held after the war took place in Strasbourg in 1920. This was a deliberate choice. Stockholm, the intended venue for what should have been an ICM in 1916, was passed over so that Strasbourg, which had been annexed by Germany after the Franco-Prussian war of 1870–71 and reclaimed by the French at the Treaty of Versailles along with the rest of Alsace and Lorraine, could be chosen. To make sure that the point was made thoroughly, the French organisers, led by Picard, made sure that German mathematicians were not invited.

In this inevitably highly charged atmosphere individual actions were remembered. Klein had signed the notorious Declaration to the Cultural World, in which 93 German scientists and professors had affirmed that they stood foursquare behind the Kaiser and the war.[1] Hilbert had not been so nationalistic. In 1917 he defied protests from German students and published an obituary of Gaston Darboux, the distinguished French geometer. Hilbert's personal stock was therefore high abroad, and he was well placed to play a leading role in bringing Germany back into the international fold and assuaging enmities. This he proceeded to do.

He was not alone. The mission to the Gulf of Guinea off West Africa to observe the solar eclipse, which Eddington led in May 1919, verified Einstein's remarkable predictions about the ability of gravity to bend light. This was hailed not only as a startling new insight into the nature of space, but as a reconciliation between countries only recently at war. The circumstances of the mission were complicated, the necessary observations delicate and fraught with error, and it was not until 6 November that a joint meeting of the Royal Society and the Royal Astronomical Society in London confirmed that Newton's theory had finally to be modified in favour of Einstein's. The *Times* reported the story the next day, and a huge swell of public excitement took it up. The fact that Einstein was German was more than offset by the fact that he had not signed the notorious Declaration, and Eddington, a Quaker and therefore a pacifist, could write that 'For scientific relations between England and Germany this is the best thing that could have happened.'[2] The same spirit of reconciliation was manifest in Einstein's response to the British and American enthusiasm, and undoubtedly helped dampen patriotic zeal for attributing blame and seeking retribution.

When people and funds began to return to the German universities, Göttingen managed to prosper again, and with Klein

---

[1] Cassidy (1992) p. 562, n. 21, notes that the letter was then circulated as a petition which eventually 4000 German professors signed and only 3 did not, of whom one was Einstein.

[2] Quoted in Fölsing (1993) p. 445.

in illustrious retirement the leader of the prestigious Mathematics Institute was now Hilbert alone. The mathematical world he could survey bore great scars. A number of young German mathematicians had been killed in the War, but many had survived, not least because Germany had had a policy of finding military uses for such specialists that tended to keep them away from the front line. In France there had been no such policy, and many young French mathematicians had to be counted in the long lists of the War dead. Some estimates put the numbers of students of mathematics and science who enrolled in 1914 and were then either killed or wounded at over 40%.[3] Jean Dieudonné, who helped re-build French mathematics between the wars, wrote in 1986 that the number of young mathematicians able to conduct research and lead the next generation of students had been reduced almost to nothing.[4] Mathematics in France was throughout the 1920s in the hands of those who had been too old to fight. Its strengths lay in its tradition, exemplified by the remarkable figure of Jacques Hadamard, whose seminar was for many years the focal point of mathematical research in France, but insofar as innovation comes most easily to the young, France was necessarily at a disadvantage.

Although the destruction was less marked in Britain, and much less in America, the taste for mathematics was less secure than it was in Germany. In England, which still largely meant Cambridge, Baker had shifted his interests from complex function theory to Italian-style algebraic geometry. He produced a string of gifted geometers, who did well at collecting the prestigious Smith's Prizes, but only a few had the strength to escape the problems inherent in what became a rather self-indulgent tradition and make truly original contributions. Arguably, the most vital thing Baker did was institute the Saturday afternoon tea parties, which were in effect high-powered research seminars, the first of their kind at Cambridge. Much more important, ultimately, was the reign of Hardy and Littlewood, who established a powerful modern school of analysis.

---

[3] Beaulieu, unpublished.

[4] Dieudonné is quoted by Beaulieu, unpublished.

In the United States, the end of the war was a period of mixed fortunes. Mathematics at Harvard was in transition from teaching to research. A number of new professors were appointed, including Kellogg (a former student of Hilbert's and an expert in potential theory) Walsh, and Marston Morse (who led the world in applying topological ideas to the study of manifolds). This gradually gave Harvard an edge that was being blunted elsewhere. Chicago in the 1920s has been characterised as a PhD mill, Yale was much less energetic. Princeton too was a place where its staff were burdened with heavy teaching loads. MIT, where Wiener worked, continued to grow steadily in the 1930s, but elsewhere the depression that started in 1929 naturally put an end to any hard-won improvements in the 1920s, and despite some remarkable achievements the two decades from 1919 to 1939 are not spoken of warmly by the mathematicians of the time.

When Hilbert looked East, the situation seemed at the mercy of politics. In the new Soviet Union mathematicians were more or less left alone, but the country was poor and the circumstances for research difficult. Moreover, at times the regime cracked down on travel abroad, leaving Soviet mathematicians isolated. The countries of the former Austro-Hungarian Empire took a variety of different paths. In Poland a carefully thought-out but risky plan to develop mathematics along a few overlapping lines was implemented, with a remarkable degree of success. Polish mathematicians became famous for their work in real analysis, topology, and mathematical logic. Hungary, which had built up a tradition in mathematics before the war and had an energetic policy of fostering mathematical talent in the young, was able to continue doing so until the Fascist regime of Horthy came to power and many mathematicians and physicists emigrated. The most visible sign of this was the remarkable number of Hungarians who worked on the atomic bomb project during World War Two.

1925 is a good year to take stock of the post-war situation. Göttingen was largely restored, and was undoubtedly the leading university for mathematics in Germany. Hilbert was surrounded by a new group of powerful mathematicians: in algebra Emmy Noether; in analysis Courant and Herglotz; in number theory

Landau (brought from Berlin) and Siegel; in foundations of mathematics Bernays. It was natural for young, gifted mathematicians growing up in Central Europe to think of visiting Göttingen, and that is what happened. The wide-ranging and quick-witted von Neumann had arrived from Hungary in 1924 and spent a lot of time with Hilbert. However, in that year Hilbert fell ill with pernicious anaemia, at the time a degenerative and ultimately fatal disease. His doctors gave him only months to live, but by chance his friends came to hear of an experimental method of treatment being pioneered by G.R. Minot at Harvard. According to Reid,[5] Courant sent Minot a long telegram urging him to help, and enlisted Oliver Kellogg, a former student of Hilbert's, now at Harvard, to keep up the pressure. Kellogg drew in others, notably George Birkhoff. Minot protested that the treatment was still at an early stage, and required large amounts of raw liver from which only small amounts of the effective ingredient could be extracted, but eventually he relented and gave instructions for the preparation of raw liver that would keep Hilbert alive until he, Minot, could make enough of the medicine to send to Göttingen. This was done, and not only was Hilbert's life saved but he was restored to something like his former self. His survival is eloquent testimony to the very high regard in which he was held internationally.

In 1926 physics was transformed utterly and forever by the arrival of quantum mechanics. What could have destroyed Hilbert's reputation in applied mathematics was paradoxically to save it, although it pushed his 6th Problem firmly into the shade. Much of the most important work was done at Göttingen by a small group of people in their early twenties; the jocular term at the time was 'boy physics'.[6] The impression was underlined by the silence of the older physicists, men such as Planck and von Laue of the waning Berlin school. In July 1925 Heisenberg (Born's assistant at Göttingen and then 23) published his first paper on the subject, which describe a prodigious set of transformation rules for measurable quantities in physical processes. Two months later Born

---

[5] Reid (1970) p. 180.

[6] *Knabenphysik*, Pais (1988) p. 251.

(who was all of 37) and Pascual Jordan (also at Göttingen and then 22) observed that the transformation rules are those of matrix multiplication, admittedly for matrices with infinitely many entries, and derived the equation that Heisenberg was to base his uncertainty principle upon in March 1927. The same formula was derived independently by Dirac (also 22) in Cambridge and published in November 1925. Also in November of that year Born, Heisenberg, and Jordan published a joint paper on the foundations of the new subject, which they called matrix mechanics. In January 1926 Pauli (25) derived the discrete hydrogen spectrum by matrix mechanics. In January 1926 Schrödinger (also 37) published an alternative formulation ('wave mechanics') which he used to obtain the continuous hydrogen spectrum, and in June 1926 Born published the first probabilistic interpretation of the theory. This succession of remarkable events may be said to end (as Pais says it does) with Bohr (age 32) and his principle of complementarity, proclaimed in September 1927.[7]

The new physics was confusing even to its discoverers, on both mathematical and physical grounds. Einstein's opinion of Heisenberg's first paper, expressed in a letter to Ehrenfest, was 'In Göttingen they believe in it (I don't).'[8] Whatever relief was felt with the arrival of Schrödinger's approach, which offered familiar partial differential equations in place of unfamiliar matrices, was tempered by the unease that there could be two such apparently dissimilar accounts of the fundamental processes of physics. Indeed, to the physicists it was a good physical interpretation that the new mathematical methods lacked, and which Born's probabilistic account began to provide.

What Heisenberg had encoded as a set of rules relating observed quantities before and after some event, Schrödinger described as the propagation of matter in the form of a wave, thus connecting the new physics with some of the best-known aspects of the old. In Heisenberg's approach, discrete behaviour (quantum jumps) came about because the matrices had a discrete set (or spectrum) of

---

[7] Pais (1988) p. 253.

[8] Quoted in Pais (1988) p. 253.

eigenvalues. According to Schrödinger, the conditions that made the propagation appropriate to any given case imposed the eigenvalues as parameters describing a basic set of solutions. For a time it seemed as if this was an entirely different approach to Heisenberg's, and the two men were at some pains to emphasise their differences. But in April 1926 Pauli circulated a letter and in May 1926 Schrödinger published a proof that the two were mathematically equivalent. Thereafter, most physicists found wave mechanics easier to work with most of the time, but intense disagreement grew up concerning the interpretation of the theory, which was gradually resolved in favour of Heisenberg.

In the heat of such advances, which was not to let up for many more years, physicists felt the lack of interpretation most keenly. For as long as any generally convincing interpretation was lacking, they could have no sympathy with an attempt to axiomatise physics that assumed the fundamentals were properly understood. But they had a corresponding respect for the mathematics that enabled them to resolve one by one many of the difficulties that had begun to plague their previous theory of atomic physics, and for this reason they acquired a renewed respect for Hilbert. This was not out of some vague awareness of the power of mathematics, but from a precise knowledge of the contents of the first volume of Courant and Hilbert's book, *Methods of Mathematical Physics*, which had come out in 1924, just before all the excitement began (and was written, we are told, entirely by Courant).[9]

This book covered the theory of matrices and linear transformations, including the infinite dimensional case, familiar to Hilbert since his work on 'Hilbert space' in 1906; the series expansion of arbitrary functions, including Fourier series, Legendre polynomials and other related systems, and the related spectral theory; linear integral equations, the calculus of variations; vibration and eigenvalue problems, including Sturm–Liouville theory and the cases of

---

[9] As Courant says in the preface: 'The responsibility for the present book rests with me. Yet the name of . . . Hilbert on the title page seems justified by the fact that much material from Hilbert's papers and lectures has been used.' In fact, Hilbert's lectures on mechanics of 1924 were more elementary, and inclined to the calculus of variations.

discrete and continuous spectra; the application of the calculus of variations to eigenvalue problems; and the study of special functions defined by eigenvalue problems.

There could hardly have been a more prescient collection of topics, down to the inclusion of infinite matrices, integral equations, and the whole theory of eigenvalues. Hilbert was delighted. Even his choice of language had turned out unexpectedly well: 'I developed my theory of infinitely many variables from purely mathematical interests,' he marvelled, 'and even called it "spectral analysis" without any presentiment that it would later find an application to the actual spectrum of physics!'[10]

In later years Heisenberg wrote that

> Indirectly Hilbert exerted the strongest influence on the development of quantum mechanics in Göttingen. This influence can be fully recognised only by one who studied in Göttingen during the twenties. Hilbert and his colleagues had created there an atmosphere of mathematics, and all the younger mathematicians were so trained in the thought processes of the Hilbert theory of integral equations and linear algebra that each project which belonged in this field could develop better in Göttingen than in any other place. It was an especially fortunate coincidence that the mathematical methods of quantum mechanics turned out to be a direct application of Hilbert's theory of integral equations.[11]

Hilbert, Brouwer, and mathematical logic

In the 1920s Hilbert's own energies were concentrated in mathematical logic. What in 1900 had seemed to be merely a challenge to provide an underpinning to the uncontentious nature of the integers, a problem that would surely yield to an analysis mounted with sufficient care and energy, rapidly became much harder than that, as the boundary between logic and mathematics began to shift and blur. Hilbert had started out with the mathematician's common-sense view of mathematics, to wit that the elementary

---

[10] Quoted in Reid (1983) p. 183.
[11] Quoted in Reid (1983), p. 182–183.

parts of it (Euclidean geometry and arithmetic) are true. Euclidean geometry was idealised physics, out of which other geometries were spun in more or less standard ways. As his appreciation of axiomatic mathematics deepened, and his view of the fundamental mathematical objects changed, the common-sense view became less tenable, and Hilbert was forced to ponder what was 'true' and what was 'proved' in his system. What, indeed, was it to *prove* something?

In 1904 he had addressed the International Congress of Mathematicians at Heidelberg on the subject of his 2nd Problem, the foundations of arithmetic. After providing a rather brisk characterisation of various older views, he proclaimed his faith in the axiomatic method. But whereas an axiomatic foundation of geometry could take arithmetic for granted, the foundations of arithmetic itself had to stand on what, exactly? Hilbert opted for logic, but with a twist. Since arithmetic necessarily involved talk about sets, and naïve set theory was plagued with paradoxes, Hilbert now wanted the simultaneous development of arithmetic and logic. He proceeded to sketch how this could be done, but the result was not very convincing, and Hilbert retreated for several years.

Even the 1904 paper represents a shift for Hilbert. The most likely interpretation of his position in 1900 was that any geometry described by (a consistent set of) axioms can be described in coordinate terms, over some field of numbers which might be rather odd, and numbers, of whatever kind, can be described in purely logical terms. By 1904 he has begun to doubt this. The obvious basis for the number concept is the concept of a set. Numbers should emerge as features of sets, as naïvely they do: we have a good sense in simple cases of the number of elements in a set. Then familiar techniques of algebra should permit mathematicians to stretch the number concept far enough to cover Hilbert's unusual geometries. But there are problems with the very concept of a set. Moreover, logical rules are going to have to be spelled out to deal with infinite sets, such as the set of all integers. Very tentatively, therefore, Hilbert speculated that one could develop number theory and logic together. This was to prove an unexpectedly arduous route, and one can doubt if Hilbert's buoyant optimism had let him suspect just how difficult a task this could be.

In 1917 Hilbert went to Zürich in neutral Switzerland and gave a lecture, later published under the title Axiomatic Thought.[12] There he gave one of his tours of mathematics, pointing out ways in which various axioms played significant roles, and showing how consistency of the axioms could be established by reducing the problem to a simpler system. This method, he then observed, broke down when it was a case of giving a foundation either to the integers or to set theory. "But since the examination of the consistency is a task that cannot be avoided, it appears necessary to axiomatise logic itself and to prove that number theory and set theory are only parts of logic." He then noted that this task had been prepared by Frege and most successfully explained by Russell; here I think 'most successfully' is to be taken in the sense of 'best yet'. Hilbert then required that the task be concluded by solving a number of difficult epistemological questions: the problem of the solvability in principle of every mathematical question, the problem of the checkability of every result; the need for a criterion of simplicity of proofs, the relationship between content and formalism in mathematics and logic, and finally the problem of the decidability of a mathematical question in a finite number of steps (can a given statement be shown to be provable/refutable in a given theory, or is it independent?). He concluded his lecture with an illustration of the nature and importance of the last of these, the decision problem.

We see here that what had begun as a set of prescriptions for specific problems in mathematics, such as Diophantine questions in number theory, has become a programme for all of mathematics. The stakes have been raised, which does not necessarily make the problem any easier, or mean that Hilbert himself had precise ideas about how his new problems could be solved. What he still had, evidently, was his optimism. He felt able to specify the terms by which anyone of good will would agree that a mathematical task had been carried out (they are very close to saying that there

---

[12] Hilbert (1917).

should be criteria, agreed in advance, by which it can be agreed if a question has an answer, and, if so, if a candidate for an answer can be checked, all in finitely many steps).

By 1922 matters had taken a darker turn. German professors had fallen to a bout of pessimism – some have argued as a result of their nation's defeat in the war. In mathematics this manifested itself in what is called the Foundational Crisis.[13] Prominent in this was Brouwer, the Dutch topologist who had long cultivated an anguished branch of the philosophy of mathematics. Hilbert felt personally challenged by the younger figure of Brouwer, whom he had once held in high regard as a topologist. But Brouwer had always held strong views on the nature of logic and language, and was deeply attracted to the idea that there is much we cannot know (and certainly much we cannot communicate). In particular, he felt that the human mind was strikingly limited in its ability to deal with infinite sets, where the logical principle of the excluded middle was, he argued, no longer applicable. When infinite sets of objects were under consideration, he argued that the usual dichotomy (a statement is either true or it is false) no longer applies. More precisely, he said that the only way to show that a set contains such-and-such an element was explicitly to construct such an element. It was not enough to prove that the assumption that the set does not contain this element leads to a contradiction. That this philosophy of intuitionism (as it was called) had severe implications for analysis was no problem for Brouwer, who, in his way, was as principled as Hilbert. But Hilbert, who had based many of his finest mathematical achievements on abstract existence proofs, such a philosophy was anathema.

What made matters worse was that for a while at least Hilbert's finest student, the most wide-ranging of all the bright young men drawn to Göttingen, Hermann Weyl, also found Brouwer's ideas attractive. Weyl too had a deep interest in philosophy, and he too was willing to strike down certain mathematical arguments on philosophical grounds.[14] Although there are real differences

---

[13] See Mehrtens (1990).

[14] In Weyl's case Fichtean idealism, see Scholz (1995).

between their positions, in the early 1920s both Brouwer and Weyl agreed that a mathematical object can only be said to exist if an explicit construction can be given for it. Naturally, Hilbert could not agree, and their philosophically minded attack on much of mathematics was intolerable to him. So he replied in 1922 with a famous paper, delivered in Hamburg.[15]

Hilbert placed his trust in signs, even to the point of amending St John's dictum to read: 'In the beginning was the sign.' He thus interpreted the single stroke, with any finite number of its repetitions, as meaningless signs. Meaningful signs, on the other hand, came in the form of abbreviations, such as 2 for 1 + 1. Hilbert also introduced a variety of additional signs for identification purposes, thus a special sign to express 'is a number', and another for 'is a function', etc. A separate category of signs was adopted for variables, along with a third category of signs for communication. He then outlined how formulae, implications, and proofs were to be written, and called for a proof theory to study the scope of such a formalised scheme.

Seen as a strategic move, Hilbert's plan was nothing less than a bold attempt to outflank the intuitionists. In the face of their claim that parts of mathematics could not be formalised, he proposed to identify all mathematical knowledge with the stock of provable formulae, reserving inferences with content for those propositions pertaining to the higher level of his new metamathematics. Everything in sight was designed to be finite, from a finite array of symbols to the proofs, all of finite length. The crux of the matter then became clear: it remained to prove, on these foundations, the validity of the principle of the excluded middle for infinitely many numbers, functions, and functions of functions. This, Hilbert argued, would give an essentially finitistic justification for logical arguments involving infinite objects. If this program succeeded, then Brouwer's intuitionism was defeated.

Hilbert's bold attempt to make the philosophy of mathematics into a (new) part of mathematics did not convince his opponents, whereas in 1927 Brouwer gave a series of successful lectures in

---

[15] The *Neubegründung* or New Grounding, Hilbert (1922).

Berlin, followed by another, more philosophical, in Vienna the next year. With youth on his side (Brouwer was a mere 46 to Hilbert's 65) his clear advocacy of German nationalism in the face of the international boycott of German scientists, as well as his formidable mathematical abilities, Brouwer must have looked to Hilbert like the coming man. Hilbert, on the other hand, was still slowly recovering from his pernicious anaemia.

Unable to win the philosophical battle, and appalled by what he saw as the consequences for mathematics of the intuitionist position, Hilbert decided that he had to protect the *Mathematische Annalen*, the journal of which he was the editor-in-chief and which he saw as the leading journal for all of mathematics. Brouwer was an editor, and had been since 1915. In October 1928, Hilbert decided to sack him. The ensuing saga was messy and unpleasant. The journal had never confronted such a problem before, and it was unclear that an editor could be sacked. Einstein, another editor, refused to be drawn into the dispute, even though he considered Brouwer to be a psychopath, and he referred to the whole business as a 'frog-and-mouse battle'.[16] Brouwer's initial response was to say that Hilbert was now of unsound mind, thus immediately alienating any supporters he might have had. Arguments raged, touching on the vexed relationship between Berlin and Göttingen, and on the ending of the boycott of German mathematicians at the Bologna ICM of 1928 (where it was generally agreed that the Hilbert camp had come out better than the nationalists led by Bieberbach, with whom Brouwer was allied). Finally Hilbert won, as he was bound to, and Brouwer took it badly. He retreated into himself, and published little on intuitionism. Since Weyl had also returned to orthodoxy, although for different reasons, this left Hilbert's philosophy of mathematics and proof theory as the leading system of ideas.

Hilbert continued, with his hard-working assistants Ackermann and Bernays, to elaborate proof theory, and in 1931 he published a forceful re-statement of his views on the occasion of his becoming an honorary citizen of his native Königsberg. In that lecture he

---

[16] All ths is enjoyably described in van Dalen (1990).

urged the unity of thought and the unity of the laws of nature with an almost mystical passion, and asked: how can we understand the coincidence of nature and thought, theory and experiment? He answered: by means of mathematics. He called up the shades of many mathematicians in his support, ending with his illustrious predecessor in Königsberg, Jacobi, who had said: 'The sole purpose of mathematics is the honour of the human spirit.' The lecture ends with Hilbert's moving affirmation of his deepest belief about mathematics: 'there are absolutely no unsolvable problems. Instead of the foolish *ignorabimus*, our answer is on the contrary:
We must know,
We shall know.'

Mathematical logic after Gödel

Instead, the fate that is popularly held to have lain in store for Hilbert's programme was the destruction of his hopes for the foundations of mathematics with the work of Gödel. In 1929 Kurt Gödel was a 23-year old Austrian logician. He had studied in Vienna, where he had been taken up by the mathematician Hans Hahn, who was interested in mathematical analysis, topology, and logic. Hahn introduced him to a group of philosophers around Moritz Schlick. This group, later known as the Vienna circle, was responsible for the philosophy of logical positivism, which was hostile to most forms of metaphysics and sought to make philosophy a science. Gödel was much more strongly inclined towards metaphysics, and gradually distanced himself from the group, but he may have been encouraged to study logic by Carnap's lectures on mathematical logic, and Carnap was a prominent member of the circle.

Gödel's first major result is known as his completeness theorem, and it forms the basis of his doctoral thesis, which was finished in the summer of 1929. It makes an important claim about a set of axioms for the propositional calculus laid down by Russell and Whitehead and similar to those used by Hilbert and Ackermann in their book of 1928. To understand it, it is necessary to separate the ideas of true and provable in one's mind, and to be clear about how we deal with mathematical statements. As every one who learns mathematics knows, mathematical statements can be thought of as just strings of symbols, which are manipulated according to formal rules. If we stick with the view that the statements are strings of symbols, then a proof is obtained by the right formal manipulation of symbols. To use a distinction that was coming into vogue only a little before Gödel's time, this is a syntactic view of mathematics. But we often like to think of the symbols as being about something (numbers, or shapes, perhaps). If we interpret the symbols so that they become meaningful, we can do something else: we can say that such-and-such a mathematical statement becomes true or false. This is a semantical view of mathematics. Gödel's completeness theorem was a semantical one: it said that the axiom system he considered was semantically complete, which means that if a sentence is such that its negation can not be interpreted as a true statement, then it can be proved. The deductive system is said to be complete because it delivers 'all' the semantic consequences of the hypotheses.

The result is technical, and in the present context it is arguably more important for the way it opened up the distinction between true and provable. Gödel argued for this distinction in a way that directly referred to Hilbert's claim that there are no unsolvable problems in mathematics. Taking 'consistent' to mean that no contradiction can be derived using only finitely many formal inferences, Gödel said that his result was equivalent to the following: Every consistent axiom system consisting of only valid formulae has a realisation. Why should one want to do this, he asked? Might it not be claimed that the mere consistency was sufficient guarantee that a model exists, in which case the proof is otiose? Such was Hilbert's view in his correspondence with Frege. Gödel noted that Brouwer had insisted the year before in Vienna that the consistency of an axiom system did not imply without further ado that a model can be constructed. Gödel and Brouwer were, however, temperamentally very far apart, and Gödel might not have been personally swayed by his philosophy. What did concern him was the idea that consistency, as a matter of mathematics, could only be taken to imply the existence of the objects whose properties were defined by the axioms if one also knew that every mathematical problem is solvable. Gödel gave this argument in support of this position. What must be pre-supposed, he said, is, that we cannot prove the unsolvability of any problem. For, 'if the unsolvability of some problem (in the domain of real numbers, say) were proved, then, from the definition [of existence] above, there would follow the existence of two non-isomorphic realizations of the axiom system for the real numbers, while on the other hand we can prove the isomorphism of any two realizations. We cannot at all exclude out of hand, however, a proof of the unsolvability of a problem if we observe that what is at issue here is only unsolvability by certain *precisely stated formal* means of inference.'[17] Gödel therefore set about refining what it means for axioms to be consistent, statements provable, and so on. Hilbert's optimism was about to be put to the test of a rigorous formulation.

Gödel established the completeness of one particular set of rules, but they did not enable one to do mathematics. Could a set of rules that describe a worthwhile piece of mathematics be proved complete? What Gödel showed in his paper 'On formally undecidable propositions'[18] was that if Pk is any consistent axiom system containing an axiom system, P, adequate to produce elementary arithmetic and perhaps some other axioms, then the statement that expresses the consistency of Pk in the symbolism of P is unprovable in Pk. In fact, as soon as Gödel had finished the proof of this theorem, he wrote: 'I wish to note expressly that [this theorem does] not contradict Hilbert's formalistic viewpoint. For this viewpoint presupposes only the existence of a consistency proof in which nothing but finitary means of proof is used, and it is conceivable that there exist finitary proofs that *cannot* be expressed in the formalism of P.'[19]

Gödel announced his incompleteness theorem at a conference in Königsberg, of all the most appropriate places, on 7 September

---

[17] Gödel (1929), in Gödel (1986) pp. 61, 63.
[18] Gödel (1931).
[19] See Gödel (1986) p. 195.

1930—by an irony of history, the day before Hilbert became an honorary citizen there. Publication of the paper followed shortly afterwards. It seems that von Neumann was quick to grasp that the consistency of elementary number theory had been shown to be unprovable, but that no less a philosopher than Carnap was for some time unsure of the implications of Gödel's discovery. Hilbert himself may well have learned of it via Bernays, who wrote to Gödel to ask for an offprint of the paper when it came out and entered into a correspondence with the young Austrian. Bernays felt that either von Neumann was right or some finitary means of proof were not formalisable in the system Gödel used (this was a possibility Gödel himself had drawn attention to) and he soon came to appreciate what Gödel had done, but others were not so speedily convinced. In particular Zermelo polemicised at length against Gödel's work (apparently to little effect).

What may be taken as Hilbert's official response came in 1939 with the second volume of his book with (more precisely, written by) Bernays, *The Foundations of Mathematics*.[20] Gödel's biographer has commented that 'Whether because of its meticulous treatment of details or because of Hilbert's implied imprimatur, the book at last seems to have stilled serious opposition to Gödel's work, at least within the community of logicians'.[21] Philosophers remained more sceptical, or simply failed to grasp the point. Dawson quotes Gödel as saying of Wittgenstein that he advanced 'a completely trivial and uninteresting misinterpretation'.[22] On the other hand, formalists were moved to deepen their understanding of formal systems, and far from making people despair of human reasoning the general response to Gödel's work has been, as the American logician Emil Post predicted, to re-affirm the creative power of human reason.[23]

---

[20] Hilbert and Bernays (1934).

[21] Dawson (1991) p. 94.

[22] The majority view on Wittgenstein as a philosopher of mathematics is probably low, but he naturally has his passionate supporters. An interesting comparison of his views with those of the mathematician Bourbaki which is less sympathetic to the traditonal approach to the foundations of mathematics will be found in Kreisel (1976).

[23] Taken from Dawson (1991) p. 97.

Box 5.1 Undecidability

In the 1930s Alonzo Church in America, Alan Turing in England, and a number of other people gave theories of what functions are computable by algorithms. The work was all done independently, but the different formulations were shown to be equivalent, which must have contributed to a feeling that it was also philosophically right. They showed that a computable function necessarily has a finite definition involving only finitely many symbols. It follows that there are only countably many of these functions, but a variant of Cantor's diagonal argument shows that there are uncountably many number-theoretic functions. It follows that there are undecidable propositions.

Finding acceptable finitary proofs has proved elusive. In 1936 Gentzen showed that the consistency of elementary number theory could be derived by using a limited form of transfinite induction, but this has commonly been felt to be going too far. The situation therefore resembles an unfinished project, and mathematicians may well have felt that prodigious efforts of mathematical logic expended on shoring up a system of elementary mathematics in which every one had complete confidence was artificial unless an agreed basis could be found. So far, that has not been the case. When Turing and Church independently showed in 1936 that the problem of the decidability of a mathematical question in a finite number of steps is also unsolvable, the original hopes for Hilbert's programme were all in tatters.

Gödel's result is known as his Incompleteness Theorem, because it asserts that the theory in question can always have consistent but unprovable statements added to it. Many people think that we still do not fully understand it. An indication of this is the extraordinary result known as Presburger's theorem (this was its author's Master's thesis, written in 1930; he died in the war in 1943). Presburger showed that a first-order theory of the integers with addition (but not all of arithmetic) is actually complete. Why the dividing line between completeness and incompleteness should run through arithmetic, rather than include it or exclude it entirely, is not at all clear.[24]

Soviet Union in the 1930s, Egorov

The situation for mathematicians in the Soviet Union was dominated by the need to work round, or in some cases with, the views of the Communist Party.[25] In the late 1920s and early 1930s the Party had made an example of the distinguished analyst D.F. Egorov for lack of political zeal. Egorov had never sympathised with the aims of the Communist regime, and he was attacked in a campaign orchestrated by Ernest Kolman, the sinister and vituperative exponent of the witch-hunts over science versus ideology. Egorov was forced to resign as President of the Moscow Mathematical Society, the Mathematical Institute in Moscow was reformed to admit more proletarian personnel, the examination system was made easier as part of a 'proletarian war' on Egorov, and finally Egorov was arrested. He died in exile in Kazan in 1931, at the age of 61.

Egorov's death did not halt the campaign, which was now turned against those who did not approve of his arrest as a counter-revolutionary, and many now found it prudent to support the campaign. By going along with it, Alexandrov became elected President of the Moscow Mathematical Society in 1932. But the zealots now turned on N.N. Lusin, a difficult person who, in the 1910s and 1920s, had been the head of a strong school of Russian analysts (naturally enough called Lusitania). Lusin was now head of the mathematical section of the Soviet Academy of Sciences, but his past betrayed strong links with foreign mathematicians and intellectual links to idealism. Kolman denounced him for his allegedly 'intuitional' approach to the natural numbers, and built up an absurd case that Lusin was sabotaging Soviet mathematics. A series of articles in Pravda during July 1936 denounced him, and a trial

---

[24] Presburger's life and work are described in Zygmunt (1991).
[25] This information is taken from Demidov (1993).

was prepared. Alexandrov, who was a former pupil of Lusin but who had long found his intense emotionalism difficult and had left Lusitania in 1922 to work on topology, supported the campaign, as did several others. Even Kolmogorov, a former star of Lusitania, made some criticisms of Lusin, but Bernstein spoke in his defence. The most courageous opponent of Kolman was the physicist Kapitsa, who wrote to Molotov about it, and the aeronautics engineer Chaplygin also protested. Then, just as the process seemed unstoppable, it was halted, presumably from above and for reasons which are still not clear, between July 11 and 13 (the Zinoviev and Kamenev trials were to take place that August). Its malign energies flowed instead back towards the philosopher Florensky, who had been arrested in 1933 and was killed in a labour camp in 1937. As a result of the Lusin affair, personal connections were highly charged and mattered greatly in the world of Moscow mathematics. Lusin made sure that Alexandrov was never elected to the Academy of Sciences until after his death in 1953. The recent reminiscences of many Soviet mathematicians make it clear that decent behaviour and a certain courage on occasions counted for a lot.

Gelfond's work on Hilbert Problem 7

At the same time as Soviet mathematicians were experiencing the malign interest of their political masters, one of their number accomplished a major, and unexpected, breakthrough which resolved a part of Hilbert's 7th Problem: to show that α^β is an irrational transcendental number when α is algebraic and not equal to 0 or 1 and β is irrational and algebraic. Carl Ludwig Siegel, the distinguished German mathematician who went on to work on this problem with some success, recorded hearing a lecture of Hilbert's in 1919 in which he said that he was optimistic that he would live to see the Riemann hypothesis proved, that perhaps the youngest members of his audience would live to see Fermat's last theorem proved, but as for the transcendence of $2^{\sqrt 2}$, no-one present would live to see that! He could hardly have been more wrong. The solution of special cases of Hilbert's problem, for numbers of the form α^β, where β is a quadratic irrational (so, for example, $2^{\sqrt 2}$)

was initially obtained by the Soviet mathematician A.O. Gelfond in 1929 by a method that used Newtonian interpolation to approximate an exponential function. He obtained the general solution in 1934 by an ingenious and delicate use of the theory of complex functions, as did Th. Schneider later the same year.

Since then the most remarkable result in this difficult area has been Alan Baker's result that the number $\beta_1 \log \alpha_1 + \ldots + \beta_n \log \alpha_n$ is transcendental if either

1. there are no rational numbers $q_1, \ldots q_n$ (not all of which are zero) such that $q_1 \log \alpha_1 + \ldots + q_n \log \alpha_n = 0$; or

2. there are no rational numbers $q_1, \ldots q_n$ (not all of which are zero) such that $q_1\beta_1 + \ldots + q_n\beta_n = 0$. Baker was awarded a Fields Medal for his work at the ICM in Nice in 1970.

Pontrjagin on Problem 5

Another success of the Soviet school followed swiftly, with Pontrjagin's work on Hilbert's 5th problem. The ideas of Sophus Lie and Wilhelm Killing were taken up and developed by Élie Cartan and Hermann Weyl, and by the 1930s it was clear that they amounted to the discovery that there were four infinite families of Lie groups and their associated algebras, and five exceptional cases. Each infinite family was indexed by a number, which can be thought of as the size of the matrices used to exhibit the elements of the group. This discovery made sense of the fact that there were indeed only four systematic ways known of writing down groups of matrices with interesting geometric properties; the five exceptional groups came as a nice surprise. Moreover, the Cartan–Weyl theory contained a rich analysis of the groups themselves, allowing many of their abstract properties to be determined. At the same time the groups were being introduced successfully into physics, by Wigner, von Neumann, and others, and their central importance was well established.

All these developments gave a renewed importance to Hilbert's 5th problem. Since the Lie groups were now all known, it was natural to re-phrase the problem so that it asked for a characterisation of Lie groups among some large class of groups. The class Hilbert had alluded to were those groups that transform figures in a space, be it Euclidean or projective, in a geometrically interesting way. The Lie groups turned out to be groups that were themselves straightforward subsets of Euclidean spaces (simply because an $n \times n$ matrix has n^2 entries, which can be thought of as its coordinates in a Euclidean space of that dimension). In the jargon of the trade, the groups were locally Euclidean. On the other hand there were large classes of groups that were not like that at all, and which presumably were not relevant to the problem, even if they did have a geometric interpretation. This suggested to mathematicians to study locally Euclidean groups, and to re-express Hilbert's 5th problem in these terms: is every locally Euclidean group a Lie group?

In this formulation, a locally Euclidean group is both a group and a certain kind of topological space. The group multiplication law can be thought of a set of maps of the group to itself, so multiplication by the element a sends g to ag, which we can write as $m_a(g) = ag$, for every a and g in the group. If this map is required to be continuous then the group is called a topological group. In particular, a locally Euclidean group is a special kind of topological group. This is the context in which Hilbert's 5th problem was treated in the 1930s, because a number of mathematicians had been developing just such a theory. This in turn was part of a broader trend to find ways in which topological ideas could simplify and clarify many branches of mathematics. This last trend had undoubtedly not been predicted by Hilbert in 1900, and the rapid development of the subject was largely done by people outside the Göttingen tradition.

In topology it was always clear that some topological spaces are easier to understand than others. Hilbert himself, in his work on the calculus of variations, observed that a space that contains the limits of all convergent sequences of points is better to study than one that does not. Two different conditions guarantee such a happy eventuality, and of these the relevant one is called compactness. This is the natural generalisation to topology of the concept of finiteness, and just as finite problems are usually easier than infinite ones, so too problems about compact spaces are easier than problems about non-compact ones. So it was very reasonable for the Soviet mathematician Pontrjagin to take up Hilbert's 5th problem for compact locally Euclidean groups, and this is what he solved in an important paper of 1934.[26]

Lev Semënovich Pontrjagin was a remarkable mathematician. He had been blinded as a result of an accident at the age of 13, but took up mathematics under the guidance of Alexandrov and soon became a world leader in the new subject of topology. Russians who knew him in the 1930s said that he knew much more than he could publish, and very little that others published around the world was not already known to him.[27] Even the important work he did on the 5th problem was almost an interlude between his studies in algebraic topology and his work on the general theory of topological groups. He started to write his book *Topological Groups* in 1935. He had high standards and the task took him two years. Solomon Lefschetz in Princeton wanted to see it published by the University Press there, and after some prudent worrying about the possible political repercussions Pontrjagin sent the book himself and it was translated. The book was a success both abroad and at home, where Pontrjagin was awarded the Stalin Prize, second class, on its account. The prize of 1000 roubles kept him and his family alive during their lengthy war-time evacuation to Kasan.

Just as compact spaces are easier to study than others, so too commutative groups are easier than non-commutative ones.[28] In 1933 von Neumann was able to solve Hilbert's 5th problem for commutative groups. Important as this and Pontrjagin's achievements were, they made essential use of the extra conditions they had imposed on the groups, and it was clear that the general problem required new methods. These were not to be discovered until the American mathematicians Montgomery and Zippin, and independently Andrew Gleason, were able to take up the problem after

---

[26] Pontrjagin (1934).

[27] Fuchs (1993).

[28] A group G is commutative if and only if $gh = hg$ for every pair of elements g and h in G.

the war, in 1952. Using an important paper of the Japanese mathematician Kenkichi Iwasawa, they showed that one of two possibilities holds true: either a locally Euclidean group is a Lie group, or instead every neighbourhood of the identity element contains a subgroup. Thus was the 5th problem solved.

In its day the solution was rightly acclaimed as a major accomplishment, and as marking significant progress in the theory of topological groups. However, among the famous problems, Hilbertian and otherwise, that mathematicians have solved, the complete solution of the 5th problem has come to occupy an anomalous role. It is sometimes argued that in fact the solution of a good mathematical problem opens up the subject, but in an interview in 1986 the distinguished French mathematician Jean-Pierre Serre said: 'Still, it is true that sometimes a theory can be killed. A well-known example is Hilbert's fifth problem . . . When I was a young topologist, that was a problem I really wanted to solve—but I could get nowhere. It was Gleason, and Montgomery–Zippin, who solved it, and their solution all but killed the problem. What else is there to find in this direction? I can only think of one question: can the group of p-adic integers act effectively on a manifold? This seems quite hard—but a solution would have no application whatsoever, as far as I can see.'[29]

The 15th Problem, on the Schubert calculus

In 1903 and 1904 Eduard Study and G. Kohn both produced examples to show that Schubert's principle of conservation of number led to errors, but then Rudolf Sturm replied that Schubert had already indicated how to rule out examples of this type.[30] In 1912, the Italian mathematician Franceso Severi gave a much more thorough geometric formulation of the problem. Even so, his method had limitations and could not deal with the complicated cases. It was van der Waerden who showed in the early 1920s how this could be done, by developing the ideas of algebraic topology that

---

[29] Chong and Leong (1986) p. 13.
[30] See Kleiman (1976) p. 449.

Lefschetz had been building up on the basis of earlier ideas of Kronecker and Poincaré. Van der Waerden showed that while many of Severi's conclusions were correct, they were unproven. With the advent of more modern methods, enumerative geometry enjoyed a further burst of attention in the 1970s and 1980s.[31]

Hilbert's 16th Problem on differential equations and the work of Dulac

In 1923 the French mathematician Henri Dulac published a long paper in which among other things he showed that a differential equation of the form $X(x, y)dy + Y(x, y)dx = 0$, where X and Y are polynomials in x and y, has a finite number of what are called limit cycles. These are solution curves which are closed curves that other solution curves either spiral towards or spiral away from. In the days of valves the famous illustration of this was the Dutch physicist van der Pol's equation for a triode, which exhibits one limit cycle (shown in Figure 1). The term had been introduced by Poincaré, and the main result in the theory, due to him and the Swedish mathematician Ivar Bendixson,[32] was that under certain restrictions there are only a finite number of limit cycles. Dulac showed that these restrictions could be removed, and that there are always only a finite number of limit cycles.

His result was taken to be a solution of the second part of Hilbert's 16th Problem, even though Dulac never mentioned Hilbert or the Problem, and explicitly stated that he knew nothing other than the work of Poincaré and Bendixson. Moreover the Problem stated by Hilbert was actually more penetrating, asking for the location of the limit cycles and their number, which Hilbert implied would be determined by the degree of the polynomials X and Y. Nonetheless Dulac's result stood until the early 1990s, when the Russian mathematician Yuri Ilyashenko and the French mathematician Écalle discovered that Dulac's proof was flawed, and

---

[31] This paragraph draws heavily on Kleiman (1976) for its historical introduction to the subject.

[32] Bendixson (1901). On Bendixson, see Gårding (1998) 109–112.

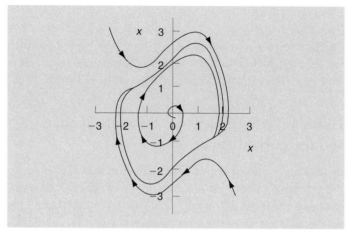

Figure 1

gave different proofs of their own.[33] Hilbert's claim that the number of limit cycles depends only on the degree of the polynomials is still unproven.[34]

Plateau problem to Douglas

An important elliptic partial differential equation that arises in a variational setting is associated with the famous Plateau problem, and although this problem was mentioned only in passing by Hilbert, later commentators, starting with Bieberbach in 1923, have usually discussed it, and so shall we.

The Plateau problem is named after the Belgian physicist Joseph Plateau (1801–83) who studied it experimentally in the 1860s. It concerns a simple closed curve in space, called a contour, and denoted Γ, and the surfaces S which span Γ, which means that the boundary of S is Γ. The Plateau problem asks for the surface of least area that spans the given contour. Such a surface is said to be

---

[33] I have taken this information from Ilyashenko (1995).

[34] It might be the case that while each pair of polynomials of degree 8, say, has only a fixed number of limit cycles, a cunning choice of polynomials can produce arbitrarily large numbers of limit cycles.

a minimal surface. They arise naturally as soap films, and it is easy and delightful to make them using nothing more than brass wire for the contour and washing up liquid mixed with glycerine for the surface. Permanent examples can be made by stiffening the film with various fixing agents.

A significant breakthrough with the Plateau problem was achieved by Riemann and Weierstrass independently in the 1860s, when they showed how the equations for a minimal surface can be obtained from complex function theory. They wrote down formulae, in terms of an arbitrary function R , now known as the Weierstrass–Enneper equations, which represent a minimal surface. However, their work does not solve the Plateau problem, where one is required to fit the surface to a prescribed contour. It proved almost impossible to deduce a suitable function R given a contour in advance. It is analogous to the situation in the elementary calculus, when you can differentiate but cannot integrate, and you wish to find the function which, when differentiated, gives you a prescribed function.

Although a physicist can think of conditions where it is natural for a contour to span a minimal surface, the mathematician can reasonably ask if it is true that every curve in space bounds a minimal surface. Could it be the case, for example, that there is a contour with the property that every surface which spans it has infinite area? It was not until 1918 that it was shown that every polygonal curve spans a minimal surface. Decisive progress only came with the arrival of the Hungarian mathematician Tibor Radó.

In a paper of 1925, Radó took up the problem from the perspective of Hilbert's problem on the analytic characteristic of solutions to certain types of partial differential equation, as solved by Bernstein. In 1930 he showed that the Plateau problem can be solved for any curve which has finite length. In a second paper Radó solved the Plateau problem for any contour which can be spanned by a disc-like surface of finite area, which is a remarkable achievement going way beyond what had earlier been achieved.[35]

Only now, on a trip to America, did Radó hear that an American

---

[35] Radó (1930a) and (1930b) respectively.

Box 5.2 Three integrals

Douglas was very clear in his paper (1933) that there are three integrals one might consider:

The Dirichlet or energy integral: $D(\phi) := \iint \left\{ \left(\dfrac{\partial \phi}{\partial x}\right)^2 + \left(\dfrac{\partial \phi}{\partial y}\right)^2 \right\} \mathrm{d}x\mathrm{d}y$,

integrated over a given region. The function ϕ is any C^2 function taken assigned values on the boundary of the region. A minimising function, if it exists, will be harmonic. It will define a surface in $\mathbf{R}^3$ spanned by the contour Γ.

The least area integral: $S(\Sigma) := \iint (EG - F^2)^{1/2} \, \mathrm{d}u\mathrm{d}v$. The surface is any C^2 surface with the prescribed contour as boundary. A minimising surface, if it exists, will have zero mean curvature.

A new integral A of his own devising. It is defined on any parametric representation of the boundary. A minimising parameterisation, if it exists, will be spanned by a surface satisfying the Weierstrass-Enneper equations, which Douglas regarded as making it of least area in the sense of Riemann and Weierstrass (although this is not obvious).

Douglas then showed that the first two integrals are difficult to study because even when it is clear that the functional has a minimum value, there need not be a function (or a surface) at which it takes that value. However, the integral he introduced was much better behaved, and using ideas of the French mathematician Fréchet, Douglas could show that his functional is bounded and attains its bounds. This gives him a preferred parameterisation, $\mathbf{g}^*$, for the contour. With this parameterisation he used standard methods in the theory of harmonic functions to obtain the equation of a minimal surface.

mathematician, only two years younger than him (33 to Radó's 35), was claiming to have done as much, or even more. This was Jesse Douglas, then an Assistant Professor at MIT. Douglas had published a number of short notes claiming a variety of impressive results, but revealing little of his methods. This may be because he had had a very hostile reception when he talked on this material in Göttingen on a trip round Europe in early 1930.[36] But starting in 1931 he published more and more, and the full originality of his approach and its remarkable extent became visible.

Douglas showed that Hilbert's direct method does not work when the Plateau problem is formulated in the usual way, but it does work when the problem is formulated his way. One gets a sense of the power and originality of Douglas's approach by observing that he showed that there is a minimising parameterisation even if his integral is always infinite. This means that the contour is spanned by a minimal surface (in Douglas's sense) even if it is spanned by no surface of finite area at all, and Douglas proceeded to give examples of contours which spanned no surface of finite area. At one extreme are contours defined by curves analytic at every point except one and yet they only span surfaces of infinite area. At the other extreme are contours so crinkly that no piece of one could be part of the boundary of a surface of finite area. Such curves have the property that you could take any point of the boundary, any arc of the boundary through that point, complete the boundary how you liked to form a closed Jordan curve, and any surface spanning the curve so formed would have infinite area. Douglas pointed out that any curve drawn wholly within the surface spanning the original curve spanned a surface of finite area. As he put it: all the area was out at the boundary. It followed that the minimal surface could not be extended beyond the original curve.

Then in a remarkable step that no-one had thought of before, Douglas looked at the case when the contour lies in a plane. His methods showed that the map from the disc to the plane region spanned by the contour given by Douglas's theorem is conformal. The conclusion is that when $n = 2$ Douglas's theorem reduces to the Riemann–Carathéodory theorem. Douglas was justifiably proud of the fact that no-one had thought to investigate the Plateau

---

[36] According to Constance Reid's biography of Courant (p. 174) and Norman Schaumberger, private communication.

problem when $n = 2$ and that when you did his method gave a new proof of one of the deepest theorems in complex function theory.

Seen dispassionately, the verdict must be that Radó was the first to solve the Plateau problem in anything like generality, but Douglas was the first to solve it in complete generality. Radó, who plainly felt challenged by the arrival of the American, could, and did, argue that the original Plateau problem made no sense unless at least one surface spanning the contour had finite area, and that while Douglas's method was more general, it was also more complicated. Radó rightly claimed that he had theorems about when branch points could not occur, and theorems giving conditions which ensured that the minimal surface was unique. None of this could be found in Douglas's work. Even so, when the first Fields Medals were presented at the International Congress of Mathematicians of 1936 in Oslo, the medal committee chose Douglas over Radó.

Bieberbach on Hilbert, 1930

In September 1930, Bieberbach was in Königsberg as part of the celebrations connected with Hilbert becoming an honorary citizen of his native city. Bieberbach gave a lecture to the Society of German Scientists and Physicians on the influence of Hilbert's Problems. He reminded his audience of the way intuition can mislead mathematicians, and promote in them a desire for rigorous proofs. But purely abstract thought can also mislead. It can seem, he said, as if one has used Beelzebub to exorcise the Devil. However, Hilbert's axiomatic approach seems to avoid one having to make such a choice, or rather, to combine the virtues of logical rigour with its own intuitive elegance. This ideal harmony of intuition and thought[37] which Hilbert proposed in 1900 has become generally accepted today, said Bieberbach, and this marks a momentous change in mathematics over the last 30 years.

Having established his general argument, Bieberbach now turned to the individual problems themselves. Among his opin-

---

[37] The German words are *Anschauung* and *Denken*, which have a fine Kantian ring to them.

ions, and he had certainly done his homework, some stand out. He was sympathetic to the idea of axiomatising physics, and noted that axiomatic theories were different things in mathematics and physics. Among the axiomatic theories of the mathematical kind that had been proposed since 1900, he noted von Mises's axiomatic account of probability theory (given in 1919 without, however, referring to Hilbert). Hilbert's student Hamel had axiomatised classical mechanics, Carathéodory had axiomatised special relativity and thermodynamics, and Hilbert, Nordheim and von Neumann had recently attempted to anchor quantum mechanics in an axiomatic system.[38]

The considerable degree of progress on the problems on number theory, much of it at Göttingen, came in for generous mention, but other Problems had fared less well. Problem 10 he interpreted in probably the only positive way one could in 1930, by alluding to the work of Siegel and Thue on the question of when polynomial equations in two variables have infinitely many solutions. He particularly lauded Siegel's surprisingly simple theorem of 1929 which showed that the genus of the corresponding Riemann surface played a decisive role. Problem 13 he called the most unfortunate of all 23 because it had only been looked at by Hilbert himself and Ostrowski.

He continued even-handedly in this way to the end. Even his own solution to part of the 18th Problem merited only a line. Then he stepped back and observed that most of these problems had proved to contain the kernel of new mathematical theories, with problems enough to interest a second generation of mathematicians. Of course, not all of the Problems were Hilbert's creation, and several domains of mathematics (Bieberbach cited function theory, topology, and differential geometry) had developed independently of Hilbert. But, he concluded, even these had clearly been affected by his axiomatic style, and the spirit and manner of his 23 Problems.

---

[38] For a recent account of how von Neumann's work fits into Hilbert's programme to axiomatise physics, see Lacki (2000).

1933–45

The terrible events of 1933 to 1945 need not be described here. A number of accurate accounts have been published of one or another part of the wholly destructive impact of the Nazis on mathematics in Germany, of which the fullest to date is Segal's.[39] Bieberbach took the opportunity to rise to the head of German mathematics, aided by a number of zealots, of whom Teichmüller stands out for his ferocity and his mathematical brilliance. Jews were expelled from Göttingen, Teichmüller himself played an ugly part in preventing Landau from continuing to give lectures, and eventually Landau made his way to Palestine. Emmy Noether fled to America, where she died unexpectedly following what should have been a routine operation. Hermann Weyl (whose wife was Jewish) also left and took up a position at the newly founded Institute for Advanced Study in Princeton. Bernays returned to Switzerland in 1934. A number of mathematicians who were not Jews emigrated in protest, Emil Artin and Carl Ludwig Siegel among them. Even Courant finally found himself in exile in New York.

Some mathematicians stayed. Helmut Hasse, the number theorist, who was a conservative German nationalist but not, in 1933 anyway, a member of the Nazi Party, took over at the Institute in Göttingen. He was certainly amongst the most distinguished of the mathematicians who were left, and it was also hoped that his politics would be acceptable to the new regime. In the event, managing mathematics was not something enthusiasts for Hitler spent much time thinking about. Partly this was because the Nazi style was not to legislate everything from the top but to invite groups to demonstrate their zeal, reckoning correctly that in a growing atmosphere of intimidation people would be only too good at doing just that. Partly this was because mathematics is not something politicians feel any great need to dabble in. Bieberbach's opportunistic posturings, his ugly account of racial types among mathematicians, his new journal *Deutsche Mathematik* that was supposed to present a suitably German way of thinking about mathematics, did not, in

---

[39] Segal (to appear).

the end, get him further than a position of influence within the German Mathematical Society.

As the years passed, from 1933 to 1939, the supply of visitors to Göttingen lessened. Deprived of the benefits of hindsight there were always those, in Germany and abroad, who believed that the worst was over, that the regime had good and bad points, that keeping in contact was better than breaking off contact, that this or that Professor was innocent of wrong-doing and would be helped by staying in touch. Plans for travel to and from Germany were made right up to the start of the War. But the reality was that Göttingen had been killed, German mathematical life had been trashed, and nothing could be done until the Nazis were defeated. Until that time, the most influential of Hilbert's followers, and very many of those who had known him personally, were living abroad, some in Great Britain, more in the United States.

Hilbert himself lived on until 1943, when he died on February 14. Sommerfeld and Herglotz attended the funeral, and Herglotz read a memorial tribute to Hilbert that Carathéodory had written but been unable to travel to deliver in person. A few months later his oldest student, Blumenthal, who had tried to survive the war in Holland, was arrested by the Gestapo. He was sent to the concentration camp in Theresienstadt, where he died at the end of 1944. Early in 1945 Königsberg was captured by Russians and razed to the ground.

CHAPTER 6

After 1945

After the Second World War, Hilbert's legacy could be viewed in many ways by his successors, and these different perceptions affected both the posthumous status of Hilbert himself and attitudes to the Hilbert Problems. The first concern was tied to his work as a profoundly creative pure mathematician, which is perhaps most intimately related to the Hilbert Problems themselves. Second came the implications he, and others, wished his work to have for applied mathematics. Third was his work on logic and the foundations of mathematics, with which he had been most actively involved in the final years of his career.

The Princeton Conference on Problems of Mathematics

A valuable perspective on the state of mathematics after the War is provided by the report of a conference held at Princeton in December 1946 and attended by 93 mathematicians, mostly American or resident in America.[1] The anonymous editors stated in their Foreword that 'it has been nearly fifty years since much thought has been broadly given to a unified viewpoint in mathematics.' After this tacit nod to Hilbert, they went on to apologise for the omission of some topics with a remark that was to prove prophetic: 'Applied mathematics, because of its wide ramifications into many sciences, could not, we thought, be treated as one field; at any event, we would be concerned with its unifying spirit, pure mathematics. Yet, as the summary shows, applied mathematics still

---

[1] Princeton Conference (1946).

makes its vitalising contributions.' The relative status of pure and applied mathematics was to prove contentious for many years.

The report then set out the discussions in several fields, and it becomes clear that the exercise was animated by a desire to put the profession back together again after the war by reminding every-one of what was known and what could most profitably be done. In so doing the contributors were not always successful; some major recent results were not mentioned, while some problems claimed to be important were to prove less significant or even, in some cases, trivial. The unifying spirit of pure mathematics is strongly felt, although it is the interesting differences of opinion that come through most clearly. There was, for example, a nice exchange between Lefschetz, the Head of the Department at Princeton, and Garrett Birkhoff[2]: Solomon Lefschetz asserted that 'To me algebraic geometry is algebra with a kick. All too often algebra seems to lack direction to specific problems,' to which Birkhoff replied 'If the algebraic geometers are so ambitious, why don't they do something about the real field?' It was then observed that 'An instance of our ignorance is Hilbert's sixteenth problem about nested ovals, still unsolved.'

Other Hilbert problems were alluded to. The still unsolved 5th Problem came up in the discussions of topology, although the discussion also made clear indirectly how much this topic had exceeded Hilbert's vision. The discussion of mathematical logic was focused on the decision problem, a Hilbertian theme that was now anchored in the work of Turing and Church. In other areas of mathematics there were topics that Hilbert had passed over, most notably various aspects of the theory of functions of several variables, and topics the speakers regarded as new. Placed after these new topics came one that, in middle age, had perhaps become respectable. Reviewing the discussion in 1988, J.L. Doob wrote: '. . . in 1946 to most mathematicians mathematical probability was to mathematics as black marketing was to marketing, that is, probability was a source of interesting mathematics but examination of the background context was undesirable.' Perhaps unaware of

---

[2]   Garrett Birkhoff was the son of G.D. Birkhoff mentioned earlier.

Hilbert's interest in the topic, and its connection to the 6th Problem, Doob then described the Russian mathematician Kolmogorov's axiomatization of probability theory in terms of measure theory (a sophisticated theory of integration and related topics). Doob himself worked with great success to bring about a unification of probability theory and measure theory, so that its practitioners may speak the language either of probability or of rigorous analysis, depending on which they are more comfortable with, and in either case develop the subject independently of its applications. As he observed, this makes both theoretical and practical work easier.

More than anything else, the Princeton conference shows two things. One is the coherence of pure mathematics, the desire of many mathematicians to draw upon a mixture of analysis, algebra, geometry and topology. This division into specialisms, each difficult to master, was, and is, felt to be artificial, even though transcending it was not thought to be easy. Some older mathematicians even doubted if it could be done any longer. The other visible feature is that as early as 1946 mathematicians were re-asserting the identification of mathematics with pure mathematics. It is to this collective preference that we now turn.

Pure or applied mathematics?

A useful perspective on the development of mathematics from 1945 to about 1980 is to see it as a struggle for the 'soul' of mathematics fought out between protagonists of pure and applied mathematics, with mathematical logicians off to one side. In that period, the enthusiasts for pure mathematics and the early Hilbert seemed to do better that those who preferred applications. Since 1980, however, the tide has turned somewhat, and mathematical logic has also come to occupy a more central position.

These debates were to be played out on a far larger stage than the close-knit environment of pre-war Göttingen. The political and economic situation dictated that there would now be an American era in mathematics. When mathematicians who had been engaged in war work as well as those who had been fighting began to return to universities, American mathematicians in particular found they had a choice: to stay close to the industrial and technological applications of mathematics that had been pioneered as part of the war effort, or to take up topics in purer branches of the subject, often not much changed since they had been set aside some years before. Would they develop a rarefied pure mathematics or a nitty-gritty applied mathematics? How would the balance be struck, and where?

Applied mathematics after the war

Courant, who had been an active member of the Applied Mathematics Panel from its inception in 1942, was naturally familiar with the world of contract work for the Navy, and used his base at New York University (NYU) to push for the continuation of such activity in peace time. The Office of Naval Research (ONR) was set up in 1946, and Courant pushed to get mathematicians already working on contracts from the ONR into permanent positions at NYU. This gave the Institute for Mathematics and Mechanics he ran there a markedly applied flavour (amusingly, it was housed for a time in premises rented from the American Bible Society). For example, the work he did with Friedrichs, and from 1946 with Cathleen Morawetz on shock waves, derived from his and Friedrichs's wartime work on bomb design. Courant set about steadily recruiting people to form the basis of the leading American community of analysts and applied mathematicians. One way of strengthening this community that was discussed around 1950 was an offer to move the entire department to Berkeley, where physics was much stronger than it was in New York (a consequence of the wartime work at Los Alamos) but Courant turned it down. In 1952 NYU was selected as the first university to have its own large, high-speed computer, and research blossomed. Shortly after he retired, in 1958, the Institute was renamed the Courant Institute of Mathematical Sciences in honour of its founder.

Applied mathematicians enjoyed propitious circumstances at the end of the War. They had good contacts with industry and politicians, and access to private and military funding. The National Science Foundation (NSF) was established in 1950, and received a great boost when the Russians launched Sputnik in 1957. That the Russians had beaten the Americans into space was a challenge to national pride and also a plausible military threat: considerable funds were unlocked to promote science in America. Yet, despite all this, many applied mathematicians came to feel that they had lost control of the mathematical enterprise to purer folk. There were complaints that 'Throughout the United States applied mathematics is given short shrift.'[3] In the 1970s mathematics in general began to enter a period of austerity, as funding from the NSF and the military declined. Peter Lax at the Courant Institute, a prominent advocate of applied mathematics, argued in 1986 that 'the separation of mathematics into pure and applied is a recent— and transitory—phenomenon,' and that 'the bold proposal to cut the lifeline between mathematics and the physical world was put forth only in the 20th century.'[4] Echoing the thoughts of John von Neumann, Lax went on,

> Besides being wrong headed, this raises profound philosophical problems about value judgements in mathematics. The question 'What is good mathematics?' becomes a matter of a priori aesthetic judgement, and mathematics becomes an art form. There is of course, some truth in this, but it seems to me that as art mathematics resembles most closely painting. In both there is a tension between two tasks; in painting, to present the shapes and colours of the visible world, and also to make pleasing patterns on a two-dimensional canvas; in mathematics, to study the laws of nature, and also to spin beautiful deductive patterns. The most successful creations are those where the tension between these tendencies is greatest; the least satisfactory are those works where one aspect predominates, as in genre painting or pure abstraction . . . . This predilection for the abstract might very well have been a rebellion against the great tradition in the United States for the practical and pragmatic; the post-war vogue for Abstract Expressionism was another such rebellion.

---

[3]   Keller, in Reid (1976) p. 276.

[4]   Lax (1986).

Lax went on to speak of a counter-tradition of 'excessive purity' and of what he called the depressing passage where Hardy exulted in his *A Mathematicians Apology* at the 'uselessness of mathematics'. The intensity of Lax's feelings on the matter might be measured by noting that it can be doubted if 'the great tradition in the United States for the practical and pragmatic' had ever embraced applied mathematics. The great names in American mathematics, and many of the lesser ones, were true to their roots and their education in a tradition derived from Germany, and as such inclined to pure mathematics. Men like E.H. Moore and Leonard Dickson who dominated Chicago, Osgood at Harvard, and others carried on that tradition when it became an autonomous one, and were in due course receptive to the émigrés from Nazi Germany. The best exceptions to this rule were G.D. Birkhoff at Harvard and Norbert Wiener at MIT. Birkhoff extended the work of Poincaré on dynamical systems and celestial mechanics. In the 1930s he worked with Koopman and von Neumann on versions of the so-called ergodic theorem, which describes how random a typical flow can be. Wiener's work had a marked probabilistic flavour too, and emphasised the study of diffusion (as in heat flow) and random motions (in his explanation of Brownian motion). But it nonetheless has a markedly theoretical cast to it. Overall, mathematics in American universities in the 1930s meant pure mathematics, and applied mathematics was done elsewhere, in departments of physics, engineering, or technology. This gave it a different flavour and a different orientation, away from theory and directly towards applications. Some pure mathematicians preferred things that way. Veblen, for example, wrote in 1929: 'I do not believe that there is, properly speaking, such a thing as applied mathematics. There is a British illusion to that effect. But there is such a thing as physics in which mathematics is used freely as a tool. There is also engineering, chemistry, economics, etc., in which mathematics plays a similar role ...'.[5]

---

[5]   Quoted in Reingold (1988).

Pure mathematics after the war

The inclination of mathematicians in the United States towards pure mathematics was strengthened by the émigrés from Germany, Italy and Hungary, who were often to claim a profound identification with the Hilbertian enterprise as they understood it. A prominent centre for pure mathematics was the Institute for Advanced Study at Princeton (IAS) which had been founded in 1933 and where Einstein was to spend the last years of his life, from 1933 to 1955. Here and at the nearby University were gathered Hermann Weyl, Emil Artin, and Kurt Gödel, as well as Lefschetz and other American notables like Veblen and Alexander. The Institute had been set up by Abraham Flexner to provide a haven for research. Staff were recruited from the most eminent researchers either to visit or, less often, to stay, and those with permanent appointments there had no teaching duties at all. Opinion remains divided as to whether this has been quite the success intended,[6] but certainly the place has been a source of much new mathematics. The old guard were no longer so productive, but they invited a group of young mathematicians who were extremely creative. In the early 1950s, for example, Lefschetz, who had made his name in the 1920s with a series of profound papers on algebraic topology, and Veblen saw to it that the young Japanese mathematician Kunihiko Kodaira was there, along with Donald Spencer, and European visitors included Atiyah, Hirzebruch, Armand Borel, and Serre. They proceeded to write several chapters of modern algebraic topology and algebraic geometry and on their return to Europe to promote it there. Modern mathematical logic was represented by Alonzo Church and even more austerely by Gödel.

The University of Chicago had been the first major centre for research in mathematics in America, but it had not prospered in the 1920s and 30s. In 1946, however, it appointed Marshall Stone as chairman of the Department of Mathematics. He had been a student of Birkhoff's at Harvard, where he had gone on to do work on the theory of operators on Hilbert space and on other branches

---

[6] Some of these debates are recorded in Regis's *Who got Einstein's Office?*

of analysis. Stone set about rebuilding a major mathematical establishment (so successfully that his reign is, naturally enough, referred to as the Stone age). He brought in the Chinese mathematician S.S. Chern, the leading differential geometer of his generation, from Princeton, Antoni Zygmund, who was to dominate the study of Fourier analysis, and Saunders Mac Lane, one of the leaders of the second generation in modern abstract algebra. He also attracted André Weil, the French mathematician who was one of the great number theorists and algebraic geometers of the period. All four of these men saw mathematics in very pure terms.

Weil and Mac Lane in particular were able to articulate a powerfully attractive vision of the mathematics enterprise, one that was in many ways a lineal descendant of Hilbert's. Or, more precisely, of a carefully selected part of Hilbert's. Similar views were held by many of the leading mathematicians at Harvard and Berkeley. These mathematicians perceived a need to direct mathematics; it should not simply be left to drift. They could easily endorse the mixture of programmes and problems that Hilbert had identified, substituting new programmes and problems of their own just as Hilbert had said every age did. They could accept that mathematics drew inspiration from the world of external phenomena. But they sympathised much more with Hilbert's talk of what happens as the human mind becomes conscious of its independence. Hilbert had indicated that problems in number theory, Galois's theory of equations, algebraic invariants and much else arose in this way. He had given new prominence to the role of conjecture in mathematics, re-shaping it so that the attention of the whole mathematical community was directed to the importance of certain unsolved problems. Generalisation and specialisation were promoted as specific ways in which theory and example could advance together, now one and now the other leading. All this is as true of applied mathematics as pure, but the detailed focus on particular problems, characteristic of much applied mathematics, lends itself less to the proclamation of a unity of mathematics. Pure mathematics, by contrast, can be portrayed as a remarkable blend of topics, and pure mathematicians were not slow to present it as such.

In 1950, the first International Congress of Mathematicians to be held after the war took place in Cambridge Massachusetts. The ICMs rapidly became a focus for international collaboration of exactly the kind that Hilbert would have welcomed, except for the unfortunate and long-running friction between the United States and the Soviet Union. McCarthyist hysteria in America meant that communists and those suspected of it found it difficult to obtain visas to visit the country. Two Fields Medals were to be issued at the Congress, one to Atle Selberg for his work on the zeros of the Riemann zeta function, a problem that Hilbert had mentioned and which was still harbouring its secrets (it still does). The other was to go to the French mathematician Laurent Schwartz for his profound reformulation and extension of functional analysis. But he was a former Trotskyist (and has indeed remained on the French left) and so he was denied a visa. The American Mathematical Society took up his case, and President Truman intervened personally to grant him a visa some months before the meeting. More bizarrely, the 85-year old Jacques Hadamard, who was not a Communist but had visited the Soviet Union several times, most recently in 1945, was also denied a visa, and he was to be one of the honorary Presidents of the Congress. The threat of a boycott by 16 of the 28 French delegates galvanised the American mathematicians and President Truman was prevailed upon to issue the visa with only five days to go.[7]

The ICMs invite experts to explain the best of their current work succinctly to rooms and even lecture halls full of other experts. Some choose to demonstrate their virtuosity, others to set out a programme. Some take the importance of their work for granted, others attempt to persuade. In general mathematicians prefer deeds to words, and to publish theorems rather than statements of intent. But some combine all these virtues, none more so perhaps than the most unusual mathematician of the 20th Century, Nicolas Bourbaki (who became a particular target of Lax's criticisms as a result).

---

[7] See Maz'ya and Shaposhnikova (1998), p. 271.

Nicolas Bourbaki on the international stage

The Frenchman Nicolas Bourbaki rose to maturity in the immediate post-war years. He was arguably the torch-bearer for a
Hilbertian philosophy of mathematics in his generation, and a
very powerful influence on and for pure mathematics. He had
begun to publish in the 1930s, driven initially by a desire to produce a usable, accurate, up-to-date book on analysis. Gradually
his aspirations broadened as his awareness of the imperfections of
existing material deepened, but his career was heavily interrupted
by the war, and when it resumed Bourbaki had the ambition to
write nothing less than an encyclopaedia of modern mathematics,
branch by branch. The initial focus on analysis had given way to
a much more aggressively pure stance: topology, commutative
algebra, integration theory. To prosecute his ideas, Bourbaki
wrote papers, attended meetings, ran a regular seminar which
Hadamard among others attended, and even announced his
birthdays. But when he applied to join the American
Mathematical Society, he was refused.[8] The reason was that
Bourbaki was not a person but a collective—a group of nine
young French mathematicians.

Bourbaki the group took his name from a French general active
in the Franco-Prussian War, and his sense of humour from the
Ecole Normale, where most of them had trained. They provided
for their succession by a strict rule that members retire at 50, and
they took their collective task of writing a coherent reference work
on mathematics very seriously indeed, meeting three times a year
for exhaustive discussions that produced endless drafts and revi-
sions, but also a steady number of volumes. These resembled
Euclid's *Elements* in many ways. Their rigour and the degree of sys-
tematisation enabled them to replace many piecemeal accounts in
the literature. Emil Artin, writing about Bourbaki's *Algebra* said
'Our time is witnessing the creation of a monumental work.'
They also recalled the *Elements* in the tedium that reading the work
at length can induce. Lorch compared reading Bourbaki with

---

[8]   See Lorch (1989) p. 149.

chewing hay.[9] But even so he strongly supported what Bourbaki was trying to do. 'Here was a movement which in my mind had been of such inestimable value in uprooting the stuffy leftovers of nineteenth century mathematics, . . .' and he described the choice he felt mathematicians had to take as eventually amounting to a separation between those 'who became a Bourbakist and [those] who on the other side [were] doomed to wander about in the once flourishing oases of the previous century.'

Several individual members of Bourbaki were even more eloquent than Bourbaki was himself. André Weil, Jean Dieudonné, Claude Chevalley and Henri Cartan, and among their successors Jean-Pierre Serre, Samuel Eilenberg, and Alexandre Grothendieck were among the most inventive mathematicians of the twentieth century, and Weil in particular shared Hilbert's desire to lead.

They also shared a philosophy of mathematics in which Hilbert was the central and indeed paradigm figure, although it was a Hilbert adapted to their purposes more than perhaps they knew. It was the axiomatic presentation of mathematics that they were keenest to attribute to Hilbert, and which they felt had been most productive for mathematics. Writing in 1939, when war with Nazi Germany seemed increasingly inevitable, Dieudonné stated his, and surely Bourbaki's, position this way.[10]

'The use of the axiomatic method, by showing clearly the source of each proposition and by showing which were the essential hypotheses and the superfluous hypotheses, has revealed unsuspected analogies and permitted extended generalizations; the origin of the modern developments of algebra, topology and group theory is to be found only in the employment of axiomatic methods.' The axiomatic method enabled previous theories to be revised, starting, Dieudonné said somewhat questionably, with Hilbert's *Grundlagen der Geometrie*, in which 'he formulated a system of 21 axioms and showed that these axioms were necessary and sufficient to prove rigorously all of the known propositions of two- and three-dimensional Euclidean geometry.'

---

[9] Lorch (1989) p. 155.

[10] Dieudonné (1939).

After reviewing the subsequent developments in mathematics, and taking a path that led through Hilbert's formalist meta-mathematical theories, Dieudonné came to the conclusion that Hilbert's axiomatic and formalist method was not only productive for technical mathematics but that its principal merit was that it brought clarity (that signal virtue in French intellectual life before deconstructionism) to mathematical thought, and final knowledge of the nature of that science. This in turn allowed it to advance science in general, for, echoing Jacobi 'The goal of science is the honour of the human spirit.' Of course, in echoing Jacobi, Dieudonné was also re-affirming Hilbert's endorsement of this belief in his address at Königsberg in 1930.

After the War, Bourbaki's views had moved from simple axiomatisation to a more elaborate philosophy of mathematics as a set of structures. The mature Bourbaki contributed to a book edited in 1947 by François Le Lionnais, and set forth a manifesto in the *American Mathematical Monthly* in 1950.[11] This richly stimulating book has its own history, for it was conceived by Le Lionnais when he was in the deportation camp Dora in 1944–45. The second edition carries Bourbaki's essay from 1950 entitled 'The Architecture of Mathematics', and was introduced to the readers as being by one of Jean Dieudonné's oldest and most faithful friends; Jean Dieudonné himself wrote on 'David Hilbert (1862–1943)', and André Weil wrote on 'The Future of Mathematics'. In fact by 1962 Dieudonné and Weil had retired from Bourbaki, but they remained close to the group. As if this was not enough, the second edition of the book reprinted Dieudonné's essay of 1939 just described.

Bourbaki's view now was that the least interesting part of mathematics was its logical formalism, important though that was in writing correct proofs. Rather, they saw as the essential goal 'precisely what logical formalism alone cannot supply: the deep lying intelligibility of mathematics'. This statement gave a novel twist to the general view of mathematicians. Mathematicians naturally believe that their subject makes sense. Unlike the captive students

---

[11] Translated into English as Le Lionnais (1971).

of unenlightened teachers, they do not give pride of place in mathematics to grinding out proofs. What mathematicians admire most in each other's work is the element of discovery, and the way in which the new results fit in with, and make sense of, the old results. They prefer arguments that explain to ones that merely demonstrate, because they are easier to use and therefore more powerful. Bourbaki's novelty was to highlight the way in which theories fit together, and to single out the axiomatic formulation of theory as facilitating this process. Bourbaki argued that it was the way mathematics is generated from a small number of axiomatic theories of different kinds that gave it its coherence, and the structure of these theories formed the architecture of the subject and made it intelligible. He saw the axiomatic conception as a guide to the entire mathematical universe. Critics of Bourbaki were to reply that the universe contained rather a lot of mathematics not dreamt of in their philosophy: algebraic geometry might be there, but not differential geometry and the rich connections to physics, nor was probability theory or any strong connection to the sciences, and nor was there much interest in algorithms and the computational side of mathematics. Indeed, Bourbaki was rather ostentatiously not interested in much of what Poincaré had advocated most highly.

In this light, certain features of Dieudonné's rhapsodic account of Hilbert stand out. He wrote (p. 305): 'What strikes one from the very first in Hilbert's works is the pure beauty of their imposing architecture; they do not give the impression of superficial "elegance", resulting from calculations cleverly carried out, but a much more profound aesthetic satisfaction, which flows from the perfect harmony between the end pursued and the means employed to achieve it.' This leads to 'the fundamental principle that in mathematics, the precise *nature* of the entities studied does not matter; it is the *relations* between these entities which alone are of importance.' The axioms provide the structure, the architecture, within which the creative mathematician can work.

The mathematical architect would create empty buildings were it not for mathematical problems. Here Dieudonné endorsed Hilbert's optimism, and reminded his readers that Hilbert had ended a lecture given in 1930 (one of his last publications), with these words. '"We must know and we shall know"—a philosophic testament worthy of the man who had brought to the science such an ample harvest of new results.' Indeed, he continued, 'few mathematicians have had as much as he the fondness, so characteristic of the true mathematician, for the special problem, exact and "concrete", if one may say that, whether such a problem is related or not to a well-developed general theory.' Dieudonné then alluded to several, without, as he put it 'speaking of his famous lecture at the Congress of 1900 on "Mathematical Problems", where a good part of those enumerated consisted precisely of "isolated" problems . . .'.

Dieudonné concluded:

> More than by his ingenious discoveries, it is perhaps indeed by the cast of his mind that Hilbert has had the most profound effect on the mathematical world; he has taught mathematicians to *think axiomatically,* that is to say, to strive to reduce each theory to its strictest logical schema, disentangled from contingencies of calculation. It is impossible to count the new and important results to which the application of this doctrine has led, and which have assured its triumph; but, more than by its immediate usefulness, it can be said that it is by its aesthetic and, even in some way, moral attraction that it has won over most young mathematicians; by his intense need to *understand,* his more and more exacting intellectual integrity, by his untiring aspiration toward a science more and more unified, pure and stripped-down, Hilbert truly personified, for the generation between the two World Wars, the ideal of the mathematician.

(I have quoted at such length from this essay because it is surely as eloquent about Bourbaki as it is about Hilbert).

Weil's essay was more closely modelled on Hilbert's presentation of 1900. Like Dieudonné, he argued that there was much more to mathematics than logical correctness.

> But, if logic is the hygiene of the mathematician, it is not his source of food; the great problems furnish the daily bread on which he thrives. 'A branch of science is full of life,' said Hilbert, 'as long as it offers an abundance of problems; a lack of problems is a sign of death.' They are certainly not lacking in our mathematics; and the present time might not be ill chosen for drawing up a list, as Hilbert did in the famous lecture from which we have just quoted.

Weil was by then well known for a set of deep conjectures that applied the methods of algebraic topology to the topic of algebraic number theory, a subject that Weil had given a strongly geometric formulation. The Weil conjectures were among the most prominent in mathematics, and were a particular source of interest to Bourbaki. Instead, Weil gave this vivid account of how the Hilbert problems then stood.

> Even among those of Hilbert, there are still several which stand out as distant, although not inaccessible, goals which will continue to suggest research for perhaps more than a generation; an example is furnished by his fifth problem, on Lie groups. The Riemann hypothesis, after the attempts to prove it by function-theoretic methods had been given up, appears to-day in a new light, which shows it to be closely connected with the conjecture of Artin on the L-functions, thus making these two problems two aspects of the same arithmetico-algebraic question, in which the simultaneous study of all the cyclotomic extensions of a given number field will undoubtedly play a decisive role. Gaussian arithmetic was centred around the law of quadratic reciprocity; we know now that this law is only a first example, we might better say the pattern of, the laws of 'class fields', which control the abelian extensions of algebraic number fields; we know how to formulate these laws so as to make them look like a coherent set. But, pleasant as this façade may be to the eye, we do not know whether it might not hide deeper lying symmetries [and] questions on which our ignorance is almost complete and in whose study the key to the Riemann hypothesis is perhaps to be found. . . . For, however wide our generalizations of Gauss' results may be, we can hardly claim to have as yet really moved beyond them. Even in the realm of abelian extensions, we have not made any progress towards the generalization of the theorems of 'Kronecker's youth dream', the generation of class fields, whose existence is known, by means of values of analytic functions. While it has been possible, without serious difficulties, to complete Kronecker's unfinished work and to obtain the solution of this problem, in the case of imaginary quadratic fields, by means of complex multiplication, the key to the general problem, considered by Hilbert as one of the most important of modern mathematics, still escapes us, in spite of the conjectures of Hilbert himself and the efforts of his pupils.

Weil then passed under review a remarkable range of current problems in mathematics, reminiscent indeed of Hilbert's presentation of 1900, before his vision of the unity of mathematics led him to conclude that 'if mathematics is to continue to exist in the way in which it has manifested itself to its votaries until now, the technical complications with which more than one of its subjects is now studded must be superficial or of only temporary character; in the future, as in the past, the great ideas must be simplifying ideas.

Programmatic statements, like many statements of intent, are not meant to shackle the speaker, and neither the Bourbaki group nor its members exactly lived according to these strictures. Recently the historian Leo Corry has argued that Bourbaki was never in practice the structuralist he claimed to be.[12] The structuralist myth nonetheless spread because it fitted with the overall French philosophical movement of the day. Structuralism, as exemplified in the work of the anthropologist Levi-Strauss, for example, was a hugely popular movement, and it was convenient for many outside mathematics to locate that obscure and often intimidating subject within it. So the educationalist Piaget could describe the child's route through elementary mathematics in structuralist terms which he naturally attributed to Bourbaki. But a more important reason for Bourbaki's influence than his insistence on structure, which Corry rightly describes as inconsistent and *ad hoc*, was the sheer calibre of the work of the individual members, that persisted through all the initial generations (and beyond).

Bourbaki, taken as a group and as a set of individuals, resembled Hilbert in many ways. His encyclopaedic volumes may stand for Hilbert's lectures. He matched Hilbert in the trenchant nature of his views, in the quality of his research, and in particular in being the focus of a community. Just as anyone willing to master German could aspire to being at Göttingen in Hilbert's time, so anyone comfortable with French could hope to work in Paris, most likely at the *Institut des Hautes Études Scientifiques* (IHES) in nearby

---

[12] Corry (1992), (1996) *see* Chapter 7.

Bures-sur-Yvette, to attend the thrice-yearly Bourbaki seminar, and perhaps even join the group. Bourbaki's influence on French mathematics was consequently inestimable, but because Bourbaki was comfortably bilingual and individual members often published in English, which many of them spoke and wrote fluently, he did equally well in a the post-war mathematical world that had become heavily anglophone.

Bourbaki's Göttingen was the IHES. It was established in 1958 by Léon Motchane, and Robert Oppenheimer of the IAS in Princeton was an advisor until his death in 1967. The first two professors Motchane appointed were Dieudonné and Alexandre Grothendieck, another Bourbaki member. Numerous other Bourbaki members, past and present, were regular attenders at IHES; Cartier, another Bourbaki member, describes the spirit of Bourbaki being there.[13] Many Americans came to visit for months at a time, some continue to come regularly and a few even stayed permanently. In this way too the influence of Bourbaki was spread abroad, just as Hilbert's had been, and this was pre-eminently the case when Grothendieck was in his prime.

Alexandre Grothendieck

Grothendieck was born in 1928, the son of a one-time political ally of Lenin who went on to fight in the Spanish Civil war and to perish in Dachau in 1943.[14] Alexandre himself ended the Second World War with his mother in a detention camp in France. He became a student of Dieudonné's, did first-rate work on infinite-dimensional vector spaces, and then turned to algebraic geometry. Here he earned a Fields Medal in 1966 for work that made him one of the leading mathematicians of the century and did more than anything else to establish the importance of Bourbaki.

Algebraic geometry, as we have already indicated, began as the study of curves, surfaces, and higher-dimensional objects defined

---

[13] Cartier, quoted in Jackson (1999).
[14] On the life of Grothendieck, see Jackson (1999).

by polynomial equations. In Hilbert's time it was generally agreed that the coefficients, and the coordinates of the points on the objects, were complex numbers. The generation of Emmy Noether and her students, as well as André Weil and Oscar Zariski, moved the subject on to allow the coefficients and the coordinates to belong to arbitrary fields. Grothendieck's idea was to broaden the subject so that everything was done over a wide class of rings, of which the integers are only the simplest, if the most important, example. This would make algebraic geometry and algebraic number theory halves of the same subject, a unification that Kronecker and Hilbert may have contemplated but which no-one had had the least idea of how to accomplish.

Grothendieck, however, was a man of formidable energy. He worked 12 hours a day every day of the week to realise every detail of his vision. He always had a clear idea of what the aims were, and what needed to be done. The scope of the task was enormous, and he enlisted Dieudonné to write out many of the details. Yet more volumes were produced by one or another group of those who attended the seminar he ran at IHES for many years. One part of it, his remarkable reformulation of the Riemann–Roch theorem, was judged so important that it was written up and published jointly by two other Bourbaki members, Armand Borel at the IAS and Jean-Pierre Serre at IHES at the end of a seminar at IAS in 1958. They drew on a set of notes Grothendieck had used for some lectures, and presented themselves as editors of a text that, they felt, was too good to be left until Grothendieck got round to writing it up whenever the architecture of his immense conceptual building permitted.

To those young enough to be caught up in the excitement, or bold enough to respond to its sheer size, Grothendieck's reformulation of algebraic geometry was a marvellous thing. Using it, Grothendieck himself mopped up two of the three remaining Weil conjectures, and his ablest follower, the Belgian mathematician Pierre Deligne, did the fourth and hardest (known as the Riemann hypothesis for finite fields). Its most recent successes have probably been Falting's solution of the Mordell and other conjectures, for which he was awarded the Fields Medal in 1984, and Andrew Wiles' solution of Fermat's Last Theorem in 1995. There is no

> **Box 6.1 Rings and fields**
>
> The familiar integers are a good example of a ring. We can add, subtract, and multiply integers together to get other integers, but we cannot divide (1/2 is not an integer). A ring is a set of objects that one can add, subtract, and multiply together and only get other elements of the same set. For example, the set of all polynomials in two variables and with integer coefficients is a ring.
>
> The familiar rational numbers are a good example of a field. We can add, subtract, multiply and divide integers together to get other integers (we only disallow division by 0). A field is a set of objects that one can add, subtract, multiply and divide together and get other elements of the same set.
>
> Throughout the late nineteenth century and twentieth century mathematicians found it easier to create a general theory of fields than of rings. Grothendieck's bold proposal to make geometry work over rings rather than fields seems certain to have implications for many years yet.

doubt that it has become a permanent part of the landscape of modern mathematics. Equally, there is no doubt that is very difficult for beginning graduates to master, and in some hands it is devoid of good examples. Some distinguished mathematicians disliked it (André Weil among them, according to what may only be a widely circulated rumour). On either interpretation, though, it was impeccably pure mathematics.

A powerful connection was created early on between Harvard and IHES, because Oscar Zariski at Harvard was favourably impressed with the new work. He had been a leading figure in the previous extension and generalisation of algebraic geometry, and now he sent his students across the Atlantic to find out what was going on and bring it back alive. Among these was David Mumford, who was able to re-open the story of invariant theory and to revive and extend Hilbert's ideas in a new, and remarkably geometric, setting. Good student of Zariski that he was, Mumford could also produce eloquent examples to show how the now-classical work looked in the setting Grothendieck had created. This helped open up the subject, and once back at Harvard and with a Fields Medal to show for his work, he helped bring into existence a strong group of American algebraic geometers. The new algebraic geometry was now secure in two continents. The strong and growing school of mathematicians in Japan also picked up the subject, as did a significant group in the Soviet Union centred on Igor Shafarevich, and the revolution (for once, not too strong a word) was accomplished.

Then, quite abruptly, Grothendieck left the field. The slight but real amount of military funding Motchane had been able to secure for IHES offended him on political grounds, and in 1970 he resigned from IHES accordingly. It was the height of the anti-Vietnam War movement, and he set up a group called *Survivre* that promoted anti-military and ecological issues. This drew the support of Claude Chevalley, a founder member of Bourbaki, and of others including Mumford for a time, but only Grothendieck went all the way and actually left mathematics. This was a long and personally difficult process, and Grothendieck made it clear that he resented the way his successors had taken up the legacy he left them without, as he saw it, appreciating its spirit or giving him due credit.

Coincidentally, his departure marked the peak of Bourbaki's influence. The group continues to exist, and to hold seminars, but by the late 1990s the supply of volumes to the encyclopaedia seems to have dried up. Such was the stranglehold it exerted in France that there has been something of a backlash there, to which Bourbaki replies that he never intended to shape mathematics education in the schools. But he is blamed for the excesses of the 'new mathematics' and for placing an undue emphasis on generalities in mathematics over interesting, and comprehensible, particulars. There is talk of a 'mid-life crisis'. The most likely explanation is that, within the mathematics profession, the balance has shifted away from large-scale theory construction and towards greater eclecticism and problem solving. Hilbert, and Bourbaki after him, stressed that theory building and problem solving go together, one now leading the other. When problems are in the lead, collective work takes second place.

Matters of taste and judgement

The disagreements over pure and applied mathematics have also what might be called an underlying political aesthetic. At one extreme stood Alexandre Grothendieck, a pacifist who refused any kind of money from military sources and could only have worked on pure mathematics. Hardy, whose pleasure in the uselessness of mathematics so offended Lax, was writing at the end of his career when the Second World War had just begun, and he felt relieved that his subject had not contributed to the destruction he saw all around him. At the other extreme one might cite a number of applied mathematicians and their friends in the physics community, for example John von Neumann, Edward Teller and Stanislaus Ulam (who worked on the hydrogen bomb) and Richard Courant.

What radical pure mathematicians opposed was not so much applied mathematics *per se* as applied mathematics in the service of government. Not applied mathematics in the style of Poincaré so much as applied mathematics in the style of Klein, with his assiduous cultivation of industry and government; and Courant, as was often said, was in this respect much more a descendant of Klein than he was of Hilbert. In that context, even its heyday as it now seems, theoretical physics was finding that students were disinclined to study the subject that had brought about the bomb. Mathematics could be pure, it could be applicable, as much of it was, and it would still be attractive—but it could not be applied without seemingly being contaminated.

Less impressionable heads could cut the cake differently. Administrators of grants, for whatever reason, had a broad vision of research. The system for funding military research in universities explicitly intended mathematicians to be paid on contract, and to stay in the university positions, not to join the armed forces. It was of course not the case that every piece of applied mathematics some one did brought them a little bit closer to the CIA, although it was the case that much of the money reaching pure mathematics did so because of the Cold War. But however shallow some perceptions of the issues may have been, it was nonetheless the case that Bourbaki's priorities seemed morally cleaner than those of the Courant Institute.

The Soviet case

The situation on the other side of the Iron Curtain was oddly similar. Mathematics was perceived as an essential part of the enterprise that would enable socialism to catch up and overtake capitalism. It would help make nuclear weapons, big rockets, put men into space. But Soviet politicians were no more capable of policing this obscure part of the enterprise than their western counterparts. They could appoint their own advisors from within the mathematics community, and they could control their budgets, but they could not understand the details of what these people said. An old Polish joke catches the situation nicely. Asked to explain why there was such a tradition of work on a topic as recondite as mathematical logic both before and after the Second World War in Poland, the mathematician replies 'Well, before the war we did it because the Church couldn't understand it, and now we do it because the Party can't understand it.' Similarly, in the Soviet Union, many talented young people were attracted to mathematics because it kept them away from the excesses of the Soviet system and allowed them to do honourable, if not very well paid, work.

After the nightmarish period of the 1930s, the war against the Nazis drew every one in, and naturally applied mathematics was promoted. In particular, there was a strong Soviet school of work on the geometric theory of differential equations, picking up on neglected ideas of Poincaré's, which Lefschetz was to recommend to American mathematicians after the war as a topic they had ignored for too long (and which Hilbert had nodded towards in the second part of his 16th Problem, to no great effect). After the war, what evolved was a mix of pure and applied mathematics that was not only traditional in Russian mathematics but reminiscent of the French settling of priorities before the rise of Bourbaki. In this spirit, Pontrjagin switched fields from topological groups to optimal control theory. To be sure, gifted mathematicians might find themselves in institutions that do not seem to western eyes to be their natural home. For example, Vladimir Drinfeld, who was awarded a Fields Medal at the ICM in 1990 for his work on the Langlands conjecture and on quantum groups (which are nothing like as applicable as their name might suggest) was working at the Institute for Low Temperature Physics and Engineering in Kharkov in 1985. But pure mathematics was as cheap in the Soviet Union as it was in the United States, and it prospered there almost as well.

The Soviet regime produced a number of distinguished mathematicians, Alexandrov among them. His life-long friend Andrei Nikolaevich Kolmogorov was regarded by many as the most gifted of them all, and Kolmogorov had a particular attraction to the Hilbert problems. In 1925 Kolmogorov formulated an axiom system for a branch of logic called intuitionistic logic, which is philosophically close to what Brouwer advocated. In 1930 he worked indirectly on Hilbert's 5th Problem, looking at one of its antecedents that had drawn Lie's particular attention. In the 1960s he wrote on the concept of an algorithm, and so contributed to Hilbert's 10th Problem. And, as we shall see, he was decisive in re-opening and solving the 13th Problem.

As is already evident, Kolmogorov, like Hilbert, worked on many branches of mathematics, including applied mathematics and mathematical logic, and contributed to the Hilbert problems as they came within his current area of interest. He began as an informal student of Lusin's in 1922, when he was 19, and that summer published a result in Fourier series that brought him international renown. In 1925 he became an official student of Lusin, and began work on probability theory, which he was to be the first to set on a rigorous mathematical footing. In 1929 he met Alexandrov, and in 1935 they bought a house together which became a centre of activity for several generations of Soviet mathematicians. At the end of his life he divided his papers into two domains: mathematics and mechanics on the one hand and probability theory and information theory on the other, corresponding, as he saw it, to the distinction between deterministic and random phenomena. Yet underneath this useful distinction he perceived, Hilbert-style, a unity, and proclaimed that there was no real boundary between order and chaos.

Mathematicians often distinguish among their number those who grasp ideas quickly and those who take them in only slowly (but perhaps more critically). Alexandrov counted Kolmogorov among the fast ones (and Hilbert among the slow). He had the ability to find rapidly what was good in the unformed ideas of other people, and conversely after the initial burst of insight that broke open a problem or a topic he was happy to leave it for many others to exploit. Such was the case with his work on Hilbert's 13th problem.

Kolmogorov and Arnol'd were leading representatives of the Russian school of mathematicians interested in differential equations and celestial mechanics, and they connected it via what is called ergodic theory to such fields as random processes, another favourite topic of Kolmogorov's. Ergodic theory is a probabilistic theory that was extensively developed in the Soviet Union and shown to apply to many fields, not just dynamical systems but also the action of groups. At the same time the study of analysis, topology, and algebraic geometry and number theory also prospered. The leader in this latter field was Igor Shafarevich, who drew around him a remarkable group of people, prominent among whom were Yuri Manin and in due course his protégé, Drinfeld. The analogy which Hilbert pointed to between algebraic numbers and algebraic functions was successfully exploited and refined by these people in a number of ways, and proved to be one of the most fruitful mathematical links between the Soviet Union and the West when politics permitted.

Mathematics in the Soviet Union was a relatively safe area for a number of people in the 1950s through to the 1980s. But one group who found themselves increasingly excluded were Jews. By the time Stalin died, in 1953, it was difficult for Jews to enter the prestigious mathematical schools, and after a brief period when things improved they then steadily worsened until by the 1970s all manner of tricks were being used to keep them out.[15] Enrolment

---

[15] See Fuchs (1993) and Saul (1999).

Box 6.2 Hilbert's 13th Problem

Hilbert's 13th Problem concerns functions of several variables. How many such functions are there? Simple addition, $x + y$, is a function of two variables: $f(x, y) = x + y$. But multiplication, $x.y$, is not new, for $x.y = e^{\log x + \log y}$. And the function $x + y + z$, seemingly of three variables, is really a function of two: $x + (y + z)$. Hilbert challenged the mathematical community to find functions of three variables that cannot be written as (possibly strange-looking) functions of two. In particular, he suggested that the function which expresses the seven roots of the equation $x^7 + ax^3 + bx^2 + cx + 1 = 0$ as functions of the three coefficients a, b, c cannot be expressed by means of continuous functions of two variables. This is, he noted, arguably the first case where functions of three variables would be required, because equations of degrees 5 and 6, although not solvable by radicals, can be simplified and so shown to be solvable using functions of only two variables.

In 1920 Ostrowski, a Göttingen mathematician, proved that there are analytic functions of two variables that are not made up of finitely many smooth functions of one variable. Hilbert himself claimed in 1927 that the roots of a polynomial equation of degree 9 could be written in terms of functions of 4 variables. The problem then seems to have lapsed until Kolmogorov astonished every one in 1956 by showing that every function of several variables can be written as a sum of composites of functions of three variables. The next year, in a seminar, one of his best students, Vladimir Arnol'd, reduced this number to two. This therefore answered Hilbert's 13th problem in the negative. It also opened the door to a whole field of study, which Kolmogorov mostly left to his students and others, which starts from the observation that the functions produced by the theorem of Kolmogorov and Arnol'd are badly behaved. If one asks: can an analytic function of n variables be written somehow in terms of analytic functions of less than n variables, the answer is 'no' (by Ostrowski's theorem).

cont. p. 213

So the question is: can every function of n variables with such-and-such a property be written in terms of functions of fewer variables with this-and-that property. Since Kolmogorov's work there have been many theorems of that kind, testifying to the unexpected profundity of Hilbert's original question, even though its answer would doubtless have surprised him.

dropped to as few as 3 to 5 Jews in a class of 500 in Moscow University. A number of influential Soviet mathematicians seem to have been anti-Semitic: the number theorist I.M. Vinogradov and Pontrjagin among them (although he denied it) and after the collapse of the Soviet Union Shafarevich's name was often mentioned in this connection. Other Soviet mathematicians did their best to oppose this anti-Semitism. Kolmogorov was among those who encouraged unregistered students to sit in on their classes. Positions for students and graduates were found in various technical Institutes, and a whole evening University was created inside Moscow State University, known as the Jewish People's University; several well-known professors taught there, not without some risk to themselves.[16]

Soviet anti-Semitism was noticed in the West, and protests began to be heard (for example, when Arnol'd was refused permission to lecture at Oxford University) and increasingly voices spoke up for Soviet dissidents in the mathematical community. Relations across the Iron Curtain were difficult at many levels. For every good contact there was a stupidity somewhere else: American mathematicians found it difficult to attend the ICM in Moscow in 1966, Soviet mathematicians could not attend ICMs in Europe, even to collect their Fields Medals (in this spirit Margulis was prevented from going to Helsinki in 1978 to collect his medal for work applying ideas of ergodic theory to the theory of Lie groups). If matters improved with Gorbachev, the consequences of the collapse of communism were nonetheless extraordinary. Mathematicians had

---

[16] Much of this information is taken from Saul (1999).

always been poorly paid in the Soviet Union, but the vigorous intellectual life they enjoyed was a compensation. After 1989 they were paid even less and steadily worse, but they were marketable. A flood of Russian mathematicians headed towards Europe (France and Germany mostly, British Universities being too starved of funds to be allowed to create the jobs) and then in considerable numbers to Israel and the USA. Margulis, for example, went to Harvard in 1990 and is now at Yale. It is not unusual now to find eminent former Soviet mathematicians enjoying a good salary in an American University while sending home money to keep mathematics alive in their home country, but mathematics in the former Soviet Union has virtually collapsed, and the once-vigorous and brilliant community there is today largely dispersed.

Hilbert's 17th Problem

Hilbert himself had worked on part of Problem 17 in 1888, well before he raised it in Paris, and had shown on abstract grounds that when polynomials with real coefficients in more than one variable are considered, there are polynomials which only take positive values but which cannot be written as a sum of squares of polynomials. It is some indication of the power of abstract methods that the first explicit example of such a thing was only found in 1966, when T.S. Motzkin exhibited $f(x, y) = 1 + x^2 y^4 + x^4 y^2 - 3x^2 y^2$. Hilbert raised the question of what can happen if a polynomial is allowed to be represented as a sum of squares of rational functions (quotients of polynomials). In 1893 he showed that for polynomials in two variables, such a representation is always possible, and in 1905 Edmund Landau deduced from Hilbert's proof that in fact four polynomials suffice.

In 1926 Emil Artin and Otto Schreier took up Hilbert's 17th Problem, and solved it completely with certain restrictions. Emil Artin was born in 1898, so he belonged to the generation after Hilbert, and he was particularly influential because of his work in number theory and abstract algebra. Towards the end of his life he delivered a paper on the centenary of Hilbert's birthday, in which he discussed the Hilbert Problems, and what he said then surely illustrates the feelings he had had as a younger man. He noted that setting unsolvable problems was only too easy to do, but that Hilbert had done something much more challenging. He had proposed problems that would promote the future development of mathematics by being both very difficult and yet also solvable with great effort. Artin and Schreier solved the 17th Problem by restricting it to what they called real closed fields. These fields are defined by the condition that -1 is not a sum of squares, so the rational numbers and the real numbers are real closed fields, but the field of complex numbers is not.[17] They showed that for such fields Hilbert was right, and every positive element is a sum of squares. Artin's methods, although rigorous, were not to every one's taste, and much subsequent work for 30 years was devoted to finding other methods, from a variety of branches of mathematics. In 1967 D.W. Dubois took up the 17th problem, on the representation of a positive function as a sum of squares, and showed that there are fields K such that a positive rational function over K is not a sum of squares, thus proving that Hilbert's question must be answered in the negative unless some condition is imposed on K. Then in 1967 Pfister showed that for fields in which Hilbert's 17th problem has an affirmative answer, the necessary number of squares depends on the number of variables, but not on the function itself. Typically, if r is the bound and n the number of variables, then $r = 2^n$.

Hilbert Problems at the International Congresses of Mathematicians

It would seem that for at least 30 years pure mathematicians won the battle to define the soul of mathematics. They did so in terms that were strikingly reminiscent of Hilbert's overall philosophy: an active and articulate version of a wide variety of mathematical topics, but one that stops short of solving genuine applied problems. Some of it is applicable, and even intended to be applied, but the implication is that that will be done by others. This was also how

---

[17] Strictly, this defines a real field; it is real closed if no extension of it is real.

the physicists of his day had seen Hilbert, and Born had defended Hilbert's work on physics only after re-defining the role of mathematics in science.[18] Of course, such a sweeping characterisation is necessarily crude, and a number of mathematicians could be cited who did genuinely get their hands dirty; but it is an accurate first impression.

Since Hilbert's stock was high, the outstanding Hilbert Problems were also well regarded. It is indicative of the high status of the surviving Hilbert problems at any one time that their solutions were often announced at an ICM. Such was the case, for example, with the Japanese mathematician M. Nagata who showed in 1959, that in its fullest generality Hilbert's 14th problem has a negative solution. He then outlined his results at the ICM in 1962 in Stockholm.

The path to this result had been opened up by Oscar Zariski, who had just given a reformulation of Hilbert's 14th Problem in geometric terms that was to prove decisive. Although it will not be possible here to enter into the details of the work of Zariski and Nagata, it is possible to show why the geometric turn was so important.

Zariski took advantage of the fact that by 1953, when he wrote his paper, there was an impressive theory that permitted one to turn many questions in algebra into questions in geometry, and vice versa. Indeed Zariski was one of the leaders in this enterprise among the generation after Emmy Noether. The translation between the two subjects goes back at least as far as Descartes, who showed how curves in the plane can be described by means of equations. The familiar example of the circle serves us here. The points where the polynomial expression $x^2 + y^2 - 1$ equals 0 are the points on the circle of radius 1 and centre the origin. Zariski had developed methods for translating subtle questions in geometry into questions in algebra, which he found easier. He could work with any number of variables, and with the sets of points that are the common zeros of any number of polynomial expressions. Such sets of points he called an algebraic variety.

He interpreted Hilbert's problem as being about two algebraic

---

[18] Born (1922) p. 93.

varieties, one inside the other, and investigated how these two varieties are related. By considering algebraic functions defined on the larger variety, he was led to a connection between Hilbert's 14th problem and a conjecture about the sets of points where these functions take the value 0, and he observed that if his conjecture were to be true then the affirmative solution of Hilbert's 14th problem would follow. But he admitted that he did not know if the falsity of his conjecture would imply that Hilbert's 14th problem had a negative answer.

Zariski observed that if the smaller variety is a curve, then indeed the result followed from a famous result in the theory of curves, the Riemann–Roch theorem. The conjecture was also true under some conditions on the field k if the smaller variety was an algebraic surface, although for somewhat *ad hoc* reasons. But there was no general result available in 1953 that permitted any deductions in the general case. What, on the other hand, could go wrong? Zariski identified the key problem. It might be that the zero sets could have unexpectedly many isolated points.

The experts saw clearly enough that Zariski's reformulation was a productive way to look for counter-examples, and in 1958 the British mathematician D. Rees found a geometric counter-example to Zariski's conjecture. Then in 1959 Nagata found another counter-example to Zariski's conjecture which he could also rework algebraically to yield a counter-example to Hilbert's 14th problem.

A negative solution to an old problem and a difficult counter-example can seem unattractive, but Hilbert had argued back in 1900 that properly seen they suggest new things to discover, and it is perhaps fitting to conclude this account of Hilbert's 14th problem by observing that, in Mumford's opinion in 1976, the subject was not only exciting but that some of its most challenging problems concern special cases. Such optimism has been born out by the developments to the end of this century, and the question of finding extra conditions that give the 14th Problem a positive answer is still eagerly pursued.

Work on mathematical logic

Two more of Hilbert's problems were to be solved in the 1960s and 1970s. They belong to the field of logic and the foundations of mathematics, the third of Hilbert's three influential legacies. Hilbert and his assistants had worked throughout the 1920s and 1930s almost exclusively on the foundations of mathematics. The 1930s saw a rush of activity as Gödel, Turing and others took up and by and large dashed Hilbert's hopes. With the growing precision of mathematical logic they showed that any theory of arithmetic must either be inconsistent or incomplete, and that the decision problem was unsolvable. Or rather, the mathematical logicians established technical results, stated with care and proved with delicate methods special to the field, that could be interpreted loosely in that way. A gap opened up between the actual achievements of the experts and what they were taken to mean.

It is not possible to set up all of mathematics along the finitistic lines Hilbert had hoped for, but perhaps something nearly as good could be achieved. Hilbert's student Gerhard Gentzen, for example, did establish a consistent formulation of mathematics, but its foundations were not as intuitively acceptable as Hilbert's. It certainly seemed as if Gödel's incompleteness theorem ruled out any natural starting point for a formalised mathematics, but perhaps Hilbert's ideas could be re-interpreted, or his aims realised in some unexpected way. Because the field had become highly sophisticated, such a task would be work for a mathematical logician, not for the multi-purpose philosopher of mathematics half-heartedly interested in what it could all mean.

After the war, distinct communities could be seen. Mathematical logicians produced a steady stream of work on logics of various kinds (there were now many different types of logic around). They also engaged more and more in model theory, a branch of the subject that had not greatly engaged Hilbert but was to prove more and more vigorous. To these mathematical logicians the failure of Hilbert's programme for the philosophy of mathematics was not an issue. It was, one might say, another vindication of Hilbert's view that a negative solution could be a positive outcome. The tools needed to overthrow that programme were powerful, and there was no shortage of problems upon which to refine them. The nature of algorithms was one such topic. Another was the status of some of the most distinctive axioms of set theory. Gödel had shown in 1939 that the axiom of choice and the continuum hypothesis are consistent with the other axioms of set theory (provided the axioms for set theory are consistent, which was not known to be the case). But he had not been able to derive them as theorems in set theory. Gödel also developed ideas about the independence of the continuum hypothesis, but these were never published in his lifetime, partly because his interests turned to philosophy after the war.

The broad community of mathematicians, however, was generally not trained, even as graduates and post-doctoral students, to follow mathematical logic, and the problems the logicians solved did not usually strike them as belonging to the mainstream of mathematics. It proved to be quite some time before even a logician as eminent as Tarski got a full professorship at Berkeley.[19] If there was a collective view of mathematical logic in the 1950s among mathematicians it was that it was a legitimate specialism, although one that even a large department might go without, and that the big philosophical issues had been resolved once and for all. Not just resolved, but resolved negatively. There were no simple foundations for mathematics, so working mathematicians had better choose some that suffice (and which would surely not turn out to be inconsistent) and get on with the 'real' job. Mathematicians' interest in the foundations of mathematics declined steeply. Bourbaki wrote themselves a generous set of rules, Paul Halmos (himself an expert in algebraic logic) provided a generation of American mathematicians with a beginner's guide with the revealing title *Naïve Set Theory*, and the situation resembled the 1890s much more than the 1930s.

There were vital areas where logic interacted with mathematics, but they emphasised the problem-oriented side of logic rather than the philosophical claims about the nature of mathematical

---

[19] See Givant (1991).

knowledge. Group theorists, for example, had some fundamental questions about when one might be said to know all the elements of a group, and these questions led straight into questions about algorithms and decision theory. It turned out that Hilbert's 17th Problem led naturally to deep questions of a logical kind.[20] Point-set topology had traditionally been another area where problems turned out to involve often quite deep questions in logic, and more cases of this kind could be adduced. But the flavour was more pragmatic, less redolent of the 'Foundations crisis' that had animated Hilbert.

This polarisation of interests within the post-war mathematical community was bad for Hilbert's reputation. It seemed as if he had spent the final productive years of his life working on problems that were either solved or somehow not part of mainstream mathematics. Just as his colleagues had initially wondered why Hilbert was so interested in elementary geometry, his successors wondered what could be so valuable in studying the foundations of mathematics. It is noticeable that in the many encomia for Hilbert that adorn the post-war period, few are for his work in this one area. The honours were given to those who had refuted him (such as Gödel and Turing) or worked on other topics entirely (for example, Tarski).

As for the Hilbert problems that concerned the foundations of mathematics, some would argue that their status was similarly affected. A widely circulated anecdote[21]—which to be sure falls under the heading 'si non e vero e ben trovato'—makes the point. 'In the early 1960s' Krantz writes, 'a brash, young, and extremely brilliant Fourier analyst . . . named Paul J. Cohen (people who knew him in high school and college assure me that he was always brash and brilliant) chatted with a group of colleagues at Stanford about whether he would become more famous by solving a certain Hilbert problem or by proving that the continuum hypothesis is independent of the axiom of choice. This (informal) committee decided that the latter problem was the ticket.' Krantz adds that

---

[20] I have not been able to discuss these here, see Sinaceur (1991) and Pfister (1995).
[21] Published in Krantz (1990).

Box 6.3 Novikov and Boone on the word problem

It is very usual for a group to crop up in a mathematical problem in terms of what are called generators and relations. The
generators are denoted by letters: $a, b, c, \ldots$. The relations are
equations between them. So for example there is a group generated by the three letters a, b, c, and the relations

$$a^2 = b^3 = e, \quad aba = b^2$$

where e denotes the identity element of the group. Elements of
the group are called words in the generators, and are such
things as ab, ba, $abab^2aba$. In this example we can recognise
the group as the symmetry group of an equilateral triangle, by
pairing a with reflection in the line l and b with the rotation
clockwise through $2\pi/3$ (marked as r in the figure). The words
ab and ba define reflections in other lines of symmetry of the
figure.

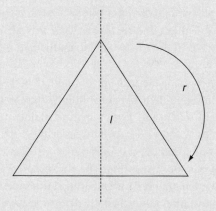

The word problem asks for a method of determining for a
given set of generators and relations which words are in fact
trivial. In the above example, $abab^2aba = e$. In the 1950s
W. W. Boone and P. S. Novikov showed that in fact the word
problem is unsolvable, in the sense that there is no algorithm
for answering it.

Cohen had been interested in this topic for years and may have conducted this seance just for fun, and then continues 'Cohen went off and learned the necessary logic and, in less than a year, had proved the independence. This is certainly one of the most amazing intellectual achievements of the twentieth century. Cohen's technique of 'forcing' has become a major tool of modern logic, and Cohen was awarded the Fields Medal for the work.'

Indeed it is true that the axiom of choice is not explicitly mentioned by Hilbert in his list of problems. But in calling for a consistent set of axioms for arithmetic, Hilbert opened the way to similar analyses of all of mathematics, and it seems reasonable to respond to his problems in their spirit as well as in the letter. Besides, by establishing the independence of the continuum hypothesis from the other axioms of set theory, Cohen did indeed settle Hilbert's first problem. There is some viability in the alternative view that, while mathematical logic always occupied a prominent place of its own at successive ICMs, the mathematical community only really sat up and took notice when the Hilbert Problems were at stake. On this view, Hilbert the mathematician was alive and well, although Hilbert the philosopher of mathematics was out of fashion.

Cohen himself made the astute observation that, despite the free use of the continuum hypothesis to simplify situations, it 'played an insignificant role in the development of mathematics, [while] the axiom of choice had been used extensively'.[22] He therefore felt that Gödel's proof that these axioms were consistent with the rest of set theory was very reassuring. The particular contribution of the continuum hypothesis is, as Hilbert had observed, to the definition of the real numbers, which are defined in terms of infinite sets of integers. Indeed, Gödel had put the problem in these terms in a paper of 1947: 'Cantor's continuum hypothesis is simply the question: How many points are there on a straight line in Euclidean space? In other words, the question is: How many different sets of integers do there exist?'[23]

---

[22] Cohen (1966) p. 85.
[23] Gödel (1947), in *Collected Works*, 2, p. 176.

Cohen located the tension between mathematics and logic one step further back[24] 'What possible objections can be raised to the construction of the real numbers? Simply this: although the reals are based on the integers, the vague notion of an arbitrary set of integers (or, equivalently, an arbitrary sequence of integers) must be introduced.' To show how this intuitive notion causes problems for a truly formal system of mathematics, he invoked the idea of a computing machine, but it is enough to imagine anyone or anything trying to follow a purely formal argument by checking every step against a book of rules. As he put it: 'If we examine Peano's axioms for the integers, we find that they are not capable of being transcribed in a form acceptable to a computing machine. This is because the crucial axiom of induction speaks about 'sets' of integers but the axioms do not give rules for forming sets nor other basic properties of sets. . . . When we do construct a formal system corresponding to Peano's axioms we shall find that the result cannot not quite live up to all our expectations. This difficulty is associated with any attempt at formalization.'

Peano's axioms do contain the notion of a set. They define the (non-negative) integers this way:

1. Every integer has a unique successor,

2. there is an integer which is not the successor of any integer,

3. two distinct integer cannot have the same successor, and

4. if M is a set of integers such that 0 is in M and such that if an integer n is in M then its successor is in M, then every integer is in M.

A mathematician, but not a logician, would say that these statements formalise the key ideas that define what we mean by the integers. The first allows you to 'add 1' to any integer. The second says you start at 0. The third says that if m and n are distinct integers, then so are $m + 1$ and $n + 1$. The last one, the axiom of induction, says that all the integers arise by taking successive successors of 0. The proofs of many formulae in mathematics

---

[24] Cohen (1966) pp. 2, 3.

accumulated over centuries depend on this principle, so it makes mathematicians uncomfortable to see it disputed. But the point, as Cohen said clearly, is to formalise something so precisely that arguments about it can be followed mechanically. Only in that way can one be really sure of their validity.

Cohen showed that a logician could make progress by replacing the vague notion of a set with a precise, if infinite, list of axioms that the integers were to obey. Indeed, one could do so in a variety of ways that captured almost the same idea. However, these formalisations could not be shown to be consistent—this is Gödel's famous incompleteness theorem. This foreshadowed the way he want on to discuss the continuum hypothesis. Now the issue was to formalise what set theory might be about and to show not only that there are formalisations consistent with the continuum hypothesis and with the axiom of choice (Gödel had already done that) but also that there are formalisations consistent with the negation of the continuum hypothesis and the negation of the axiom of choice. This shows that those axioms are independent of the axioms of set theory.

To establish the independence of the continuum hypothesis, Cohen produced a profound and difficult analysis of logical implication, the idea that some statements logically imply or 'force' others. Using his method of forcing he showed that there were models of set theory in which the negation of the continuum hypothesis and of the axiom of choice were both true. So the continuum hypothesis and the axiom of choice were independent of the usual axioms for set theory.

Cohen ended his book with some philosophical speculations about the status of the continuum hypothesis which are worth quoting at length. He rejected the view that set theory was about nonexistent fictions and so the continuum hypothesis would be neither true nor false, because such a philosophy did not explain how set theory was so successful. Moreover 'as Gödel's Incompleteness Theorem implies, we can generate statements in arithmetic provable in set theory but not in lower systems. To merely reject these statements and forever give up any possibility of deciding them is just as unsatisfactory as not knowing what to do with the continuum hypothesis.'

He then acknowledged that 'most mathematicians are more or less idealists in their view that sets actually exist and questions like the continuum hypothesis have a meaning' and rather strikingly, went on to say:

> A point of view which the author feels may eventually come to be accepted is that the continuum hypothesis is obviously false. The main reason one accepts the Axiom of Infinity is probably that we feel it absurd to think that the process of adding only one set at a time can exhaust the entire universe. This point of view regards the continuum as an incredibly rich set given to us by one bold new axiom, which can never be approached by any piecemeal process of construction. Perhaps later generations will see the problem more clearly and express themselves more eloquently.[25]

There does not seem to be any consensus on the matter even today. Perhaps there never will be, and if there is it will surely only be because some branch of mathematics has a powerful preference for one particular version of the continuum hypothesis.

Cohen was awarded the Fields Medal for this work in Moscow in 1966. In giving him the prize, Alonzo Church praised him for turning to a field not properly his own to solve a problem that has baffled the specialists.[26] He then gave a succinct indication of the nature of Cohen's work, before concluding with some remarks about the significance of it all. He observed that they struck a powerful blow against the idea that there was an absolute realm of sets, because they showed that there was not one set theory but many, whereas intuition suggests that the problem of the continuum hypothesis has a unique answer. And he knew no mathematician, he said, who believed that the continuum hypothesis was self-evident.

This point has continued to resonate throughout philosophy of mathematics. Somehow the mind has access to such objects as numbers. We all learn how to count and to conduct elementary arguments about numbers correctly. We can prove, following Euclid, that there are infinitely many prime numbers, for example.

---

[25] Cohen (1966) p. 151.
[26] Church (1966).

And if we ask how we can do this, the first, and unproblematic, steps, take us to simple rules of logic and to the idea that mathematics is founded on the primitive idea of a set. But when we start to make these precise, and inquire into the way in which laws of logic must be adapted and extended to deal with sets, and when we refine set theory to obviate the paradoxes, all that is solid really does seem to melt into air. The necessary logic is no longer simple or capable of a unique formulation. The axiomatised notion of a set is no longer intuitive and admits many distinct realisations. These are philosophical problems indeed.

Another of Hilbert's problems that looked naturally to be of a logical nature was his 10th Problem. This asks for a process which will determine in a finite number of steps whether a polynomial equation in any number of variables and with integer coefficients has any integer solutions. Particular cases were known to be solved, some answered affirmatively (the equation $x^2 + y^2 = z^2$ famously has integer solutions) and some negatively (the equation $x^3 + y^3 = z^3$ has no solutions where x, y, and z are non-zero). The novelty of Hilbert's problem is that it asks for an entirely general method which, moreover, is to be a process. Presumably he imagined some sequence of tests to which any equation could be subjected and which would classify it as either solvable in integers or not. This process neatly separates the business of showing that there are solutions to an equation from that of finding the solutions themselves. It could be entirely non-constructive, as the earliest accounts of Hilbert's basis theorem had been in his solution to the central problem in the theory of invariants, and even if it became a constructive method it might nonetheless be extremely cumbersome. Even if successful it would not put an end to Diophantine analysis and the study of the theory of numbers.

However, no general theory of finite decision processes existed in 1900, nor was one to be created until the work of Church, Post, Gödel, and Turing in the 1930s. This work only began to permeate thinking about the 10th Problem in the 1950s. In 1961 Martin Davis, Hilary Putnam, and Julia Robinson jointly wrote a paper on a related question, which the reviewer in *Mathematical Reviews* described as being only superficially related to Hilbert's 10th

Problem. As a consequence, the young Russian mathematician Yuri Matijasevich did not come across Julia Robinson's name when he began his own work on the problem, and what was to prove to be one of the most fruitful international collaborations was inadvertently postponed, for the story of the solution of the 10th Problem is a remarkable example of how shared mathematical interests transcended, or at least circumvented, the Cold War.

What the paper by Davis, Putnam, Robinson had shown in 1961 was that there is no algorithm for showing that the class of exponential equations is solvable in integers. These are equations in which the variables may also by exponentiated, for example $e^{x^2} + 3x - 1 = 0$. It might be that there is an algorithm for polynomial equations (the class considered by Hilbert) but that it breaks down for exponential equations. But suppose it could be shown that any given exponential equation has a solution if and only if a certain family of polynomial equations with coefficients determined by the exponential equation has a solution. Then Hilbert's problem is solved, and indeed solved in the negative. In an earlier paper, published in 1952, Julia Robinson had shown that this reduction can be carried out provided there is an equation in m variables and two parameters a and b which defines a relation $J(a,b)$ with properties that some experts dismissed it as wildly unnatural.

To be precise, this was the reaction of Matijasevich in the Soviet Union. Yuri Matijasevich had taken up the problem in his second year as a student as Leningrad State University in 1965. He had already obtained some new results involving some ideas in Post's work, and asked his adviser, Sergei Maslov, about where he should proceed. Maslov suggested the algorithmic unsolvability of Diophantine equations, adding 'This problem is known as Hilbert's tenth problem, but that doesn't matter to you.' To Matijasevich's protestations that he knew nothing about unsolvability of any decision problem, he replied that this was unnecessary. Unsolvability of a given problem was shown by establishing that it could be reduced to one already known to be unsolvable, and he (Matijasevich) knew the technique of reduction well

> ### Box 6.4 Julia Robinson's J-relation
>
> A relation between numbers a and b is a statement about them, such as $a^2 = b$ or $a < b$. A more substantial example is given by the relation '$R(a, b)$ if and only if $ax^2 + xy + by^2 > 0$ for all values of x and y.' This is a relation defined between two parameters a and b by an equation in two variables. It simplifies to '$R(a, b)$ if and only if $ab > 1/4$'.
>
> Julia Robinson wanted a relation $J(a, b)$ defined between two parameters a and b that was defined by an equation in m variables and with the properties:
>
> 1. for any a and b, $J(a, b)$ implies $a < b^b$, and
> 2. for any integer k there exist a and b such that $J(a, b)$ and $a > b^k$.

enough. As to what Matijasevich should read, Maslov dismissed the American work as 'most likely inadequate'.

Matijasevich's first attempts, however, proved unsuccessful, and he then read the American papers, which had been translated into Russian. He now saw that a negative solution to the 10th Problem was likely, and perhaps within reach, and he ran a seminar on the topic until the audience stopped turning up. But success eluded him, and by 1969, when he graduated and moved to the Steklov Institute in Leningrad he decided to abandon the problem. He only read the next of Robinson's papers because he was sent it to review for the Soviet counterpart of *Mathematical Reviews*, but he was at once captivated. On 3 January 1970 Matijasevich finally put all the pieces together and wrote out what he hoped was a proof of Hilbert's 10th Problem, answering it in the negative: there is no decision process of the kind Hilbert had called for. He entrusted the proof to Maslov and another mathematician for checking and left for a two-week skiing holiday with his new wife. On his return he found that they thought the proof was correct and had entrusted it to two other mathematicians, both famous for their ability to spot errors, and that they too were convinced. Matijasevich gave his first public lecture on the topic on January 29, and another mathematician in his audience then gave his account of Matijasevich's accomplishment in Novosibirsk. An American mathematician was in the audience, and he sent his notes of that talk to Julia Robinson. It turned out that these notes were very brief and most likely only someone like Julia Robinson, who had already worked very hard on the problem, could have reconstructed Matijasevich's argument from them.

Julia Robinson had studied number theory at Berkeley before writing her PhD under Tarski's supervision. It combined number theory and mathematical logic in a way that was to be characteristic of her later work. Although a long childhood illness left her with permanently impaired health, she became a leading figure in the community of mathematical logicians, and eventually was elected unopposed as the first woman President of the AMS (in 1982). She had begun to think about Hilbert's 10th Problem in 1950, and by 1970 she was in fact very close to solving it herself.

In October 1970 Julia Robinson wrote to Matijasevich. This began their correspondence, conducted without e-mail or even photocopiers. They met for the first time at a meeting in Bucharest in August 1971, and began to think of writing a joint paper, but this was difficult when communications took at least 4 weeks. Problems tended to get solved, and questions answered, before the reply came back. In the end they decided that Matijasevich would write the whole first draft and Robinson check it, and this was done, but the paper never appeared because one crucial argument involved dividing by zero. This truly elementary mistake, which every schoolchild learns to avoid, stunned Julia Robinson, who had not spotted it. She wrote to Matijasevich, in a delightfully paradoxical phrase, that 'I wanted to crawl under a rock and hide from myself.' The mistake was easy to put right, but the paper nonetheless had to be completely re-written. Their first joint paper appeared in Russian in 1974, their second, in English was largely written by Robinson.

It was now 1974, and the American Mathematical Society was busy preparing its Symposium on the Hilbert Problems. Matijasevich was invited to talk on the 10th Problem, but the Soviet regime did not grant him permission to travel, so Julia

Robinson gave the talk there. At her suggestion the paper that was published was jointly written by her, Matijasevich, and Martin Davis, and the final version was produced by Davis. There they establish the negative answer to the 10th Problem in the striking form that there is a polynomial equation P in 13 variables and depending on one parameter such that there is no algorithm which determines whether the equation has integer solutions for a given value of the parameter.[27]

Their paper makes clear that the results of the logicians had deep implications for number theory, much deeper than even Hilbert can have realised. For instance, Putnam showed that Matijasevich's main theorem implies that there is a polynomial whose non-negative values are precisely the prime numbers. This corollary was published before the main theorem was proved, and some took it to be a good reason to believe that the main theorem would be false. The best expression known to Davis and the others in 1974 needed 325 symbols and 6 lines of type.

Another problem, and one mentioned by Hilbert, that can be reduced to a Diophantine question is Goldbach's conjecture. It then becomes the assertion that a certain Diophantine equation has no solution. Had Hilbert's 10th Problem been answered positively, Goldbach's conjecture would have been answered in the negative—a connection which Hilbert surely had not suspected. Davis, Matijasevich, and Robinson showed that even the Riemann hypothesis can be reformulated as a Diophantine question (which doesn't make it any easier to solve).

More surprisingly, consider the theorems of a given formalised (but possibly inconsistent) mathematical theory. They can be recursively enumerated by writing down in some order the independent axioms defining the theory, then the deductions that can be made using just one rule of inference, then those that can be reached in two steps, and so on. Now consider the property of an integer n which says that no contradiction has been reached after n steps of the recursive enumeration of the theorems of the theory. There is a simple algorithm for this property, which is just to scan

---

[27] The number of variables has since been reduced to 9.

Box 6.5 Matijasevich's main theorem

First some definitions. Consider a polynomial equation in a certain set of variables and depending on m parameters $a_1, \ldots, a_m$: $P(a_1, \ldots, a_m; x_1, \ldots, x_n) = 0$. Think of the equation as defining the sets of m-tuples $(a_1, \ldots, a_m)$ for which the equation has a solution in natural numbers. So for example $a_1 x_1^2 + a_2 x_2^2 - a_3 x_3^2 = 0$ has solutions when $(a_1, \ldots, a_3) = (1,1,1)$ but not when $(a_1, \ldots, a_m) = (1,5,2)$; the triple $(1,1,1)$ belongs to the set defined by the equation but the triple $(1,5,2)$ does not. A set of m-tuples defined in this way is said to be a Diophantine set.

A central concept in the theoretical study of algorithms is that of recursive enumerability. A set is said to be recursively enumerable if there is a well-defined algorithm for making a list of its members. Every member of the set must occur somewhere in the list and nothing may appear in the list that is not a member of the set. For example, a Diophantine set is recursively enumerable.

Matijasevich's main theorem asserts that every recursively enumerable set of integers is a Diophantine set.

Another key concept is that of computability. A set of positive integers is computable if there is an algorithm which determines in a finite number of steps if any arbitrary positive integer belongs to the set. What makes the subject interesting is the fact that there are recursively enumerable sets that are not computable. It is this gap between recursively enumerable and computable that makes it possible to prove that some mathematical problems are algorithmically unsolvable.

the finite list of theorems looking for a pair of contradictory formulae. So the theory is consistent if and only if every positive integer n has this property. By the main theorem this says that every mathematical theory comes with (at least one) Diophantine equation which has no solutions if and only if the theory is consistent. In principle this gives a way of showing that a theory is consistent without exhibiting a model, which would have helped Hilbert avoid some criticisms of his approach to the 2nd Problem had it been available in 1900.

CHAPTER 7

Epilogue

The story of the Hilbert Problems illustrates many things about mathematics. It is does indeed demonstrate that Hilbert lifted the veil on the future. As he intended, his Problems helped in a considerable way to shape that future. They may not always have done so in ways that he expected, but they did attract several generations of mathematicians to them, the best among them. That is surely because he chose them well and gave good reasons for caring about them, and that in turn reflects his penetrating insight into the relationship between problems and theory.

Hilbert's own involvement with the foundations of mathematics changed and deepened as the twentieth century unfolded. It seems reasonable to infer that even in 1900 he imagined there could be a theory of axiom systems which could deliver such results as completeness (in his sense). There was even in 1900 a rich theory of groups, and these are structures defined by axioms. Why not a theory of other structures defined axiomatically, as Hilbert thought every branch of mathematics should be, from which would emerge a theory of axiom systems? Problem 2 asks for a completeness result for arithmetic, and Problem 3, on the equality of two volumes, suggests a sort of converse—Hilbert suggested that the axiom system in question would not provide all the results one wants. Problem 6 underscores the need for consistent sets of axioms. This is not to suggest that Hilbert had any idea of what a theory of axiom systems might be, or that he was contemplating what, in the 1920s, became his proof theory. It is merely to observe that someone who had already written a book on axioms for geometry, and wanted the axiomatic method extended, could naturally believe in 1900 that there would be a theory of axioms, without any idea of what such a theory would be.

The same confidence can be seen in his ideas about solvability, with the same deepening as the years went by. At some stage, perhaps after 1917, he began to appreciate what it can mean to ask that a problem be solvable. What in 1900 was a mathematician's view gradually became a logician's: more sophisticated, but more elusive. Take some of the Problems he considered. Problem 1, the continuum hypothesis, asks if two sets are equal. It is the kind of question that has a yes or no answer. Problem 5 suggests that the (tacit) axiom system for the theory of Lie groups is redundant. Problems 10 (the solvability of Diophantine equations) and 17 (on positive functions which are sums of squares) became markedly more logical in character than Hilbert expected, but they simply ask if something can be done. The analogy I suspect Hilbert had in mind with Problem 10 was Galois theory. Galois theory permits a mathematician to determine if any given polynomial is solvable by radicals; what is now wanted is a theory that permits a mathematician to determine if any given Diophantine is solvable in integers. Problems 19 and 20 want proofs that certain differential equations have solutions of an expected kind. Indeed, all the Problems on analysis, and its relation to physics, further illustrate the theme of how all problems are solvable, and they all were. It seems best to take Hilbert at his word. In 1900 he really believed that if mathematicians can formulate the question correctly, then they can answer it.

When it comes to number theory his taste was excellent, his touch less sure. Problem 7 on transcendental numbers was solved, against his expectation, but Problem 8, the Riemann hypothesis, continues to hold out. It is now probably the most important unsolved problem in mathematics. The call for a general law of reciprocity (Problem 9) was answered, as was Problem 11, the general theory of quadratic forms. Note, in passing, the wisdom of sticking to quadratics: as the fate of the 10th Problem suggests, there is no number theory even today of polynomials of higher degree with the richness of the theory of quadratics. Even Problem 12, poorly formulated as it was by Hilbert, was answered.

With the mixed algebraic and geometric bag that forms Problems 13–18, his taste was also less sure, presumably because he had often not worked on these topics himself. One measure of this is the way these problems held out. It does not seem as if he anchored them so securely in their theoretical context. Progress in these subjects could be made without tackling these tasks, and they languished. But even so, they got done, one way or another, in the end.

That the Problems led to major developments in the theory cannot be doubted, even if one withholds credit when the development would surely have surprised Hilbert himself. This leads up to the final point: the status of the Problems and the mathematicians who tackled them. Of course some mathematicians have had no interest in these Problems—there is a great deal else to be done. Some mathematicians' contributions have been forgotten (Ragsdale's is a case in point). Some problems have looked more exciting than others—that is only natural. But the list of those who have tackled the Problems contains many major mathematicians in the twentieth century. It is enough to cite some of the number theorists: Gelfond, Siegel, Artin, Takagi, and Hasse. The names of Dehn, Bernstein, Koebe, and Birkhoff are also illustrious, and the Problems that survived to the 1950s and 60s drew the attention of Paul Cohen, Kolmogorov, Arnold, and Zariski among others.

Almost all these mathematicians are remembered for other achievements as well, but all thought it worth their time to tackle a Hilbert Problem, and none can have begun with the guarantee of success. To ask whether they lent lustre to the Problems, or acquired lustre from the Problems, is to miss the dynamic. The Problems began with Hilbert's name, and were sustained by the prestige of Göttingen in the early years. As first Dehn, and then Bernstein, and then Koebe, and so on began to solve one or another Problem, the enterprise acquired new life. It became noteworthy to solve a Problem. Careers were assisted, reputations made. That made the Problems seem more attractive and more important. Whenever Hilbert succeeded in surrounding a Problem with the aura of enriching theory, that Problem continued to attract attention, and the unsolved Problems surely looked a little harder, a little more challenging, than the solved ones. If, unexpectedly, a theory grew up that reformulated a Problem, as was the case in logic, that Problem was likewise enhanced. Very few came to seem a little flimsy, a little beside the point.

Nor, indeed, was the solution of a Problem necessarily the end of the story. Hilbert had indicated that it should not be, that a good problem should suggest further questions. That Bernstein's work on Problem 19 gradually led to a profusion of ever more general and profound ideas only emphasised what a good Problem it had been. The few that led to dead ends were the disappointments, not the ones that led on to generations of work.

Mathematics passes through phases, if not fashions. For most of this century there has been a trend towards abstraction and structural mathematics. These are developments rightly associated with Hilbert, and there is a feeling today that we have now entered a period more sympathetic to applied mathematics, to diversity in mathematics, and to honest problem solving. But these tensions are always present. I opened this book with a quote from Minkowski, urging mathematicians 'to face problems with a minimum of blind calculation, a maximum of seeing thought'. I took it from some observations Hermann Weyl made at the Princeton Conference of 1946, but I suppressed his opinion of this view, which in many ways is a fair summary of his own approach. Weyl went on 'I find the present state of mathematics, that has arisen by going full steam ahead under this slogan, so alarming that I propose another principle: *Whenever you can settle a question by explicit construction, be not satisfied with purely existential arguments.*' (Italics were in the original.) Weyl was not reversing Minkowski's view. Most mathematicians would prefer to settle a question explicitly because the answers can be so much more useful. But it illustrates one way in which mathematicians think about their subject. Even they can find it too abstract, and perhaps, as they get older, too remote.

Mathematicians worry, as Hilbert did, that their subject is breaking up into mutually unintelligible subdivisions, a process they see happening in the sciences. They point to demographics and the sheer growth of the profession. They lament that this or that mathematician was the last to have a universal grasp: for Klein the last was Gauss and the only hope was for an organised team of mathematicians to hold the subject together. Later mathematicians have regarded Poincaré as the Gauss of his generation, or Hadamard, or Weyl, or, of course, Hilbert. Partly it depends on what branches of mathematics are to be held most closely, partly it depends on what myth, rather than accurate history, is operating. But so far, what has always happened is that branches of mathematics expand and contract. They expand as problems are solved and new ones open, as details are filled in, and new aspects of a topic are opened up. And then, when the picture gets too big, they contract as a good general theory is created so that mathematicians can re-orient themselves. There are topics that start as a slew of puzzling, loosely inter-related examples, and others that need a deep theoretical insight to get started. After a time, the puzzles need a theory, just as the insight must ultimately pay off with genuine answers. Happily, there are mathematicians who delight in detail, and mathematicians who delight in theory. Most likely the best mathematicians do both, with their own emphases and priorities.

What Hilbert saw was that there is not some facile opposition between problems (good, useful) and theory (bad, abstract) or between problems (narrow, obscurantist) and theory (good, unifying). There are good problems, and there are dull ones, and difficult problems can also be dull. There are big theories and there are small ones, but big theories can be useless. What winnows out the bad ones of either kind is a to-and-fro process whereby theories are created to solve problems and then problems bring the theories into focus. No finer illustration of this can be offered than Wiles' proof of Fermat's Last Theorem. This deservedly popular success is also a vindication of the extraordinarily abstract theory that is present-day arithmetic algebraic geometry.

Hilbert's particular skill may have lain in the ability to see, or create, the right general setting which made old, difficult problems look easy, but he created theories which worked. He may have produced some of his finest work in some of the purest areas of mathematics, but it is also worth remembering that he had a lifelong interest in physics. I would argue that Hilbert did not pose 23 problems. Nor did he urge axiomatisation and abstraction upon a concrete world. He urged the process of balancing problems with theories and developing them both together. It is in this sense, as evidenced by his 23 Problems, that he was, as Blumenthal called him, a man of problems.

HILBERT'S LECTURE
Editor's notes

1. On the text

The Euler-Mascheroni constant C is the limiting value of the quantity

$$1 + \frac{1}{2} + \frac{1}{3} + \frac{1}{4} + \cdots + \frac{1}{n} - \log_e (n)$$

as n tends to infinity. Euler found that it is approxiately 0.577218. A value obtainable on a computer package today is 0.57721566490153286061. . . . to 20 decimal places. Even today it is not known if this number is rational or irrational.

The existence of an infinite number of *prime numbers of the form* $2^n + 1$ is still not known. They arise for reasons to do with cyclotomy, see the discussion on cyclotomy p. 42.

A group G is *transitive*, or we say today acts *transitively* on a set S, if for every two points s_1 and s_2 in S there is an element g of G such that $gs_1 = s_2$. For example, the group of all rotations of a sphere acts transitively on the sphere (you can rotate the sphere so as to move any point to any other point) but the group of rotations about a fixed axis does not act transitively (no rotation about the poles will move London to Sydney).

Branching of numbers occurs when one enlarges the ring of integers. Consider the ring of Gaussian integers, which are integers of the form $m + n\sqrt{-1}$. The number 5, which is prime as an ordinary or rational integer, is a Gaussian integer ($m = 5$, $n = 0$) but it is not a prime Gaussian integer, because $5 = (2 + \sqrt{-1})(2 - \sqrt{-1})$. The prime number 5 is accordingly said to branch.

Tschirnhausen transformations reduce complicated-looking polynomial expressions to simpler ones. For example, the quintic equation $x^5 + 5ax^4 + 10bx^3 + 10cx^2 + 5dx + e$ simplifies, on setting $y = \alpha + \beta x + \gamma x^2 + \delta x^3 + \varepsilon x^4$ and eliminating x from these two expressions, to a quintic equation in y whose coefficients depend on a, b, etc and α, β, etc. It is always possible to arrange that the equation in y has the form $y^5 + 10py^2 + 5qy + e$.

The *Jacobi–Hamilton* equation (or Hamilton–Jacobi equation, as anglophones have it) is the central equation in dynamics. It is a first-order partial differential equation that expresses the equations of motions of the system.

Maurer's paper was in fact incorrect.

Newson sometimes added to Hilbert's footnotes, and I have done the same. I have labelled her footnotes MWN and mine JJG.

Otherwise I have left his footnotes unaltered, except for imposing a modern convention about italics in line with that used elsewhere in this book, and I have supplied a bibliography in the modern style, from which full references to the items he refers to can be derived.

2. On the translation

I have left the original translation almost entirely unchanged, but I have quietly changed a few words and re-ordered a sentence or two. Some technical terms have been translated differently, bringing them into line with other standard translations. Where Hilbert wrote 'Menge' and Newson put 'assemblage' I have substituted the word 'set'. Similarly, I have rendered 'Punktmenge' as 'point-set' where Newson had 'assemblages of points', 'Geschlecht' as 'genus', not 'deficiency'; 'Dreikörperproblem' as 'three body problem' not 'the problem of three bodies', 'Extremum' as 'extremal', not 'extreme'.

The term 'natural domain of rationality', for which the original German is 'naturale Rationalitätsbereich', raises two issues. The term 'Rationalitätsbereich' was introduced by Kronecker. I have translated this as a 'domain of rationality', not 'realm of rationality' as Newson did. It is a field, but one with only finitely many generators. As mathematicians active in this area after Kronecker's time pointed out, there are fields which are not domains of rationality, for example the field of all algebraic numbers, so to translate the term as 'field', which is sometimes done, can be misleading. A natural domain of rationality is either the field of rational numbers or a field of rational functions with rational coefficients in some number of variables.

Hilbert's lecture at the International Congress of Mathematicians

The Future of Mathematics

Who among us would not be glad to lift the veil behind which the future lies hidden; to cast a glance at the next advances of our science and at the secrets of its development during future centuries? What particular goals will there be toward which the leading mathematical spirits of coming generations will strive? What new methods and new facts will the new centuries disclose in the wide and rich field of mathematical thought?

History teaches the continuity of the development of science. We know that every age has its own problems, which the following age either solves or casts aside as profitless and replaces by new ones. If we would obtain an idea of the probable development of mathematical knowledge in the immediate future, we must let the unsettled questions pass before our minds and look over the problems which the science of today sets and whose solution we expect from the future. To such a review of problems the present day, lying at the meeting of the centuries, seems to me well adapted. For the close of a great epoch not only invites us to look back into the past but also directs our thoughts to the unknown future.

The deep significance of certain problems for the advance of mathematical science in general and the important role which they play in the work of the individual investigator are not to be denied. As long as a branch of science offers an abundance of problems, so long is it alive: a lack of problems foreshadows extinction or the cessation of independent development. Just as every human undertaking pursues certain objectives, so also mathematical research requires its problems. It is by the solution of problems that the strength of the investigator is hardened; he finds new methods and new outlooks, and gains a wider and freer horizon.

It is difficult and often impossible to judge the value of a problem correctly in advance; for the final award depends upon the gain which science obtains from the problem. Nevertheless, we can ask whether there are general criteria which mark a good mathematical problem. An old French mathematician said : 'A mathematical theory is not to be considered complete until you have made it so clear that you can explain it to the first man whom you meet on the street.' This clarity and ease of comprehension here insisted on for a mathematical theory, I should still more demand for a mathematical problem if it is to be perfect; for what is clear and easily comprehended attracts, the complicated repels us.

Moreover a mathematical problem should be difficult in order to entice us, yet not completely inaccessible, lest it mock our efforts. It should be to us a signpost on the tortuous paths to hidden truths, ultimately rewarding us by the pleasure in the successful solution.

The mathematicians of past centuries were accustomed to devote themselves to the solution of difficult individual problems with passionate zeal. They knew the value of difficult problems. I remind you only of the 'problem of the line of quickest descent', proposed by Johann Bernoulli. Experience teaches, Bernoulli explained in the public announcement of this problem, that lofty minds are led to strive for the advance of science by nothing more than laying before them difficult and at the same time useful problems, and he therefore hoped to earn the thanks of the mathematical world by following the example of men like Mersenne, Pascal, Fermat, Viviani and others in laying before the distinguished analysts of his time a problem by which, as a touchstone, they might test the value of their methods and measure their strength. The calculus of variations owes its origin to this problem of Bernoulli's and to similar problems.

Fermat has asserted, as is well known, that the Diophantine equation

$$x^n + y^n = z^n$$

(x, y and z integers) is unsolvable—except in certain self-evident cases. The attempt to prove this impossibility offers a striking example of the inspiring effect which such a very special and apparently unimportant problem may have upon science. For Kummer, spurred on by Fermat's problem, was led to the introduction of ideal numbers and to the discovery of the law of the unique decomposition of the numbers of a cyclotomic field into ideal prime factors—a law which today, in its generalization to any algebraic field by Dedekind and Kronecker, stands at the center of the modern theory of numbers and the significance of which extends far beyond the boundaries of number theory and into the realm of algebra and the theory of functions.

To speak of a very different region of research, I remind you of the three body problem. The fruitful methods and the far-reaching principles which Poincaré has brought into celestial mechanics and which are today recognized and applied in practical astronomy are due to the fact that he undertook to treat anew that difficult problem and to come nearer to a solution.

The two last mentioned problems—that of Fermat and the three body problem—seem to us almost like opposite poles—the former a free invention of pure reason, belonging to the region of abstract number theory, the latter forced upon us by astronomy and necessary for an understanding of the simplest fundamental phenomena of nature.

But it often also happens that the same special problem finds application in the most dissimilar branches of mathematical knowledge. So, for example, the problem of the shortest line plays a chief and historically important part in the foundations of geometry, in the theory of lines and surfaces, in mechanics and in the calculus of variations. And how convincingly has F. Klein, in his work on the icosahedron, pictured the significance of the problem of the regular polyhedra in elementary geometry, in group theory, in the theory of equations, and in that of linear differential equations.

In order to throw light on the importance of certain problems, I may also refer to Weierstrass, who spoke of it as his happy fortune that he found at the outset of his scientific career a problem as important as Jacobi's problem of inversion on which to work.

Having now recalled to mind the general importance of problems in mathematics, let us turn to the question of the sources from which this science derives its problems. Surely the first and oldest problems in every branch of mathematics stem from experience and are suggested by the world of external phenomena. Even the rules of calculation with integers must have been discovered in this fashion in a lower stage of human civilization, just as the child of today learns the application of these laws by empirical methods. The same is true of the first problems of geometry, the problems bequeathed to us by antiquity, such as the duplication of the cube, the squaring of the circle; also the oldest problems in the theory of the solution of numerical equations, in the theory of curves and the differential and integral calculus, in the calculus of variations, the theory of Fourier series, and in potential theory—to say nothing of the further abundance of problems properly belonging to mechanics, astronomy and physics.

But, in the further development of a branch of mathematics, the human mind, encouraged by the success of its solutions, becomes conscious of its independence. By means of logical combination, generalization, specialization, by separating and collecting ideas in fortunate ways—often without appreciable influence from outside—it evolves new and fruitful problems from itself alone, and then appears as the real questioner itself. Thus arose the problem of prime numbers and the other problems of number theory, Galois' theory of equations, the theory of algebraic invariants, the theory of abelian and automorphic functions; indeed almost all the nicer questions of modern arithmetic and function theory arise in this way.

In the meantime, while the creative power of pure reason is at work the outer world again comes into play, forces upon us new questions from actual experience, opens up new branches of mathematics; and while we seek to conquer these new fields of knowledge for the realm of pure thought, we often find the answers to old unsolved problems and thus at the same time advance the old theories most successfully. And it seems to me that the numerous and surprising analogies and that apparently pre-established harmony which the mathematician so often perceives in the questions, methods and ideas of the various branches of his science, have their origin in this ever-recurring interplay between thought and experience.

It remains to discuss briefly what general requirements may be justly laid down for the solution of a mathematical problem. I should say first of all, this: that it shall be possible to establish the correctness of the solution by means of a finite number of steps based upon a finite number of hypotheses which are implied in the statement of the problem and which must be exactly formulated. This requirement of logical deduction by means of a finite number of processes is simply the requirement of rigor in reasoning. Indeed the requirement of rigor, which has become a byword in mathematics, corresponds to a universal philosophical necessity of our understanding; on the other hand, only by satisfying this requirement do the thought content and the suggestiveness of the problem attain their full effect. A new problem, especially when it comes from the outer world of experience, is like a young twig, which thrives and bears fruit only when it is grafted carefully and in accordance with strict horticultural rules upon the old stem, the established achievements of our mathematical science.

It is an error to believe that rigor in the proof is the enemy of simplicity. On the contrary, we find it confirmed by numerous examples that the rigorous method is at the same time the simpler and the more easily comprehended. The very effort for rigor forces us to discover simpler methods of proof. It also frequently leads the way to methods which are more capable of development than the old methods of less rigor. Thus the theory of algebraic curves experienced a considerable simplification and attained a greater unity by means of the more rigorous function-theoretical methods and the consistent introduction of transcendental devices. Further, the proof that the four elementary arithmetical operations as well as the term by term differentiation and integration can be applied to power series, and the recognition of the utility of the power series as a result of this proof, contributed materially to the simplification of all analysis, particularly the theory of elimination and the theory of differential equations, and also the existence proofs demanded in those theories. But the most striking example of my statement is the calculus of variations. The treatment of the first and second variations of definite integrals required in part extremely complicated calculations, and the processes applied by the old mathematicians lacked the necessary rigor. Weierstrass showed us the way to a new and sure foundation of the calculus of variations. By the examples of the simple and double integral I will show briefly, at the close of my lecture, how this way leads at once to a surprising simplification of the calculus of variations. For in the demonstration of the necessary and sufficient criteria for the occurrence of a maximum and minimum, the calculation of the second variation and in part, indeed, the tiresome reasoning connected with the first variation may be completely dispensed with—to say nothing of the advance which is involved in the removal of the restriction to variations for which the differential coefficients of the function vary only slightly.

While insisting on rigor in the proof as a requirement for a perfect solution of a problem, I should like, on the other hand, to oppose the opinion that only the concepts of analysis, or even those of arithmetic alone, are susceptible of a fully rigorous treatment. This opinion, occasionally advocated by eminent men, I consider entirely erroneous. Such a one-sided interpretation of the requirement of rigor would soon lead to the ignoring of all concepts arising from geometry, mechanics and physics, to a stoppage of the flow of new material from the outside world, and finally, indeed, as a last consequence, to the rejection of the ideas of the continuum and of irrational numbers. But what an important nerve, vital to mathematical science, would be cut by rooting out geometry and mathematical physics! On the contrary, I think that wherever mathematical ideas come up, whether from the side of the theory of knowledge, or in geometry, or from the theories of natural or physical science, the problem arises for mathematics to investigate the principles underlying these ideas and to establish them upon a simple and complete system of axioms, so that the exactness of the new ideas and their applicability to deduction shall be in no respect inferior to those of the old arithmetical concepts.

To new concepts correspond, necessarily, new symbols. These we choose in such a way that they remind us of the phenomena which were the occasion for the formation of the new concepts. So geometrical figures are signs or mnemonic symbols of space intuition and are used as such by all mathematicians. Who does not always use along with the double inequality $a > b > c$ the picture of three points following one another on a straight line as the geometrical picture of the idea 'between'? Who does not make use of drawings of segments and rectangles enclosed in one another, when it is required to prove with perfect rigor a difficult theorem on the continuity of functions or the existence of points of condensation? Who could dispense with the figure of the triangle, the circle with its center, or with the cross of three perpendicular axes? Or who would give up the representation of the vector field, or the picture of a family of curves or surfaces with its envelope which plays so important a part in differential geometry, in the theory of differential equations, in the foundations of the calculus of variations, and in other purely mathematical sciences?

Arithmetical symbols are written figures and geometrical figures are drawn formulas; and no mathematician could spare these drawn formulas, any more than in calculation he could dispense with the insertion and removal of parentheses or the use of other analytical signs.

The use of geometrical symbols as a means of strict proof presupposes the exact knowledge and complete mastery of the axioms which lie at the foundation of those figures; and in order that these geometrical figures may be incorporated in the general treasure of mathematical symbols, a rigorous axiomatic investigation of their conceptual content is necessary. Just as in adding two numbers, one must place the digits under each other in the right order so that only the rules of calculation, i.e., the axioms of arithmetic, determine the correct use of the digits, so the use of geometrical symbols is determined by the axioms of geometrical concepts and their combinations.

The agreement between geometrical and arithmetical thought is shown also in that we do not habitually follow the chain of reasoning back to the axioms in arithmetical discussions, any more than in geometrical. On the contrary, especially in first attacking a problem, we apply a rapid, unconscious, not absolutely sure combination, trusting to a certain arithmetical feeling for the behavior of the arithmetical symbols, which we could dispense with as little in arithmetic as with the geometrical imagination in geometry. As an example of an arithmetical theory operating rigorously with geometrical ideas and symbols, I may mention Minkowski's work, *Geometrie der Zahlen*.[1]

Some remarks upon the difficulties which mathematical problems may offer, and the means of overcoming them, may be in place here.

If we do not succeed in solving a mathematical problem, the reason frequently consists in our failure to recognize the more general standpoint from which the problem before us appears as only a single link in a chain of related problems. After finding this standpoint, not only is this problem frequently more accessible to our investigation, but at

---

[1] Leipzig, 1896.

the same time we come into possession of a method which is applicable also to related problems. The introduction of complex paths of integration by Cauchy and of the notion of the ideals in number theory by Kummer may serve as examples. This way of finding general methods is certainly the most practical and the most secure; for he who seeks for methods without having a definite problem in mind seeks for the most part in vain.

In dealing with mathematical problems, specialization plays, as I believe, a still more important part than generalization. Perhaps in most cases where we unsuccessfully seek the answer to a question, the cause of the failure lies in the fact that problems simpler and easier than the one in hand have been either incompletely solved, or not solved at all. Everything depends, then, on finding those easier problems and on solving them by means of devices as perfect as possible and of concepts capable of generalization. This rule is one of the most important levers for overcoming mathematical difficulties ; and it seems to me that it is used almost always, though perhaps unconsciously.

Occasionally it happens that we seek the solution under insufficient hypotheses or in an incorrect sense, and for this reason do not succeed. The problem then arises: to show the impossibility of the solution under the given hypotheses, or in the sense contemplated. Such proofs of impossibility were effected by the ancients; for instance, when they showed that the ratio of the hypotenuse to the side of an isosceles right triangle is irrational. In later mathematics, the question of the impossibility of certain solutions plays a prominent part; and we perceive in this way that old and difficult problems, such as the proof of the axiom of parallels, the squaring of the circle, or the solution of equations of the fifth degree by radicals, have finally found fully satisfactory and rigorous solutions, although in another sense than that originally intended. It is probably this remarkable fact along with other philosophical reasons that gives rise to the conviction (which every mathematician shares, but which no one has as yet supported by a proof) that every definite mathematical problem must necessarily be susceptible of an exact settlement, either in the form of an actual answer to the question asked, or by the proof of the impossibility of its solution and therewith the necessary failure of all attempts. Take any definite unsolved problem, such as the question as to the irrationality of the Euler–Mascheroni constant C or the existence of an infinite number of prime numbers of the form $2^n + 1$. However unapproachable these problems may seem to us and however helpless we stand before them, we have, nevertheless, the firm conviction that their solution must follow by a finite number of purely logical processes.

Is this axiom of the solvability of every problem a peculiarity characteristic only of mathematical thought, or is it possibly a general law inherent in the nature of the mind, a belief that all questions which it asks must be answerable by it? For in other sciences also one meets old problems which have been settled in a manner most satisfactory and most useful to science by the proof of their impossibility. I cite the problem of perpetual motion. After seeking unsuccessfully for the construction of a perpetual motion machine, scientists investigated the relations which must subsist between the forces of nature if such a machine is to be impossible;[2] and this inverted question led to the discovery of the law of the conservation of energy, which, again, explained the impossibility of perpetual motion in the sense originally intended.

This conviction of the solvability of every mathematical problem is a powerful incentive to the worker. We hear within us the perpetual call: There is the problem. Seek its solution. You can find it by pure reason, for in mathematics there is no *ignorabimus*.

The supply of problems in mathematics is inexhaustible, and as soon as one problem is solved numerous others come forth in its place. Permit me in the following, tentatively as it were, to mention particular definite problems, drawn from various branches of mathematics, from the discussion of which an advancement of science may be expected.

Let us look at the principles of analysis and geometry. The most suggestive and notable achievements of the last century in this field are, as it seems to me, the arithmetical formulation of the concept of the continuum in the works of Cauchy, Bolzano and Cantor, and the discovery of non-Euclidean geometry by Gauss, Bolyai, and Lobachevsky. I therefore first direct your attention to some problems belonging to these fields.

---

[2] Königsberg, 1854. See Helmholtz, 'Ueber die Wechselwirkung der Naturkräfte und die darauf bezüglischen neuesten Ermittelungen der Physik'; Vortrag, gehalten in Königsberg, 1854.

1. Cantor's problem of the cardinal number of the continuum

Two systems, i.e., two sets of ordinary real numbers or points, are said to be (according to Cantor) equivalent or of equal *cardinal number*, if they can be brought into a relation to one another such that to every number of the one set corresponds one and only one definite number of the other. The investigations of Cantor on such point-sets suggest a very plausible theorem, which nevertheless, in spite of the most strenuous efforts, no one has succeeded in proving. This is the theorem:

Every system of infinitely many real numbers, i.e. , every infinite number- (or point-), is either equivalent to the set of natural numbers, l, 2, 3, . . . or to the set of all real numbers and therefore to the continuum, that is, to the points of a line; *as regards equivalence there are, therefore, only two sets of numbers, the countable set and the continuum.*

From this theorem it would follow at once that the continuum has the next cardinal number beyond that of the countable set; the proof of this theorem would, therefore, form a new bridge between the countable set and the continuum.

Let me mention another very remarkable statement of Cantor's which stands in the closest connection with the theorem mentioned and which, perhaps, offers the key to its proof. Any system of real numbers is said to be ordered, if for every two numbers of the system it is determined which one is the earlier and which the later, and if at the same time this determination is of such a kind that, if a is before b and b is before c, then a always comes before c. The natural arrangement of numbers of a system is defined to be that in which the smaller precedes the larger. But there are, as is easily seen, infinitely many other ways in which the numbers of a system may be arranged.

If we think of a definite arrangement of numbers and select from them a particular system of these numbers, a so-called sub-system or subset, this sub-system will also prove to be ordered. Now Cantor considers a particular kind of ordered set which he designates as a well-ordered set and which is characterized in this way, that not only in the set itself but also in every subset there exists a first number. The system of integers l, 2, 3, . . . in their natural order is evidently a well-ordered set. On the other hand the system of all real numbers, i.e., the continuum in its natural order, is evidently not well-ordered. For, if we think of the points of a segment of a straight line, with its initial point excluded, as our partial set it will have no first element.

The question now arises whether the totality of all numbers may not be arranged in another manner such that every subset may have a first element, i.e., whether the continuum cannot be considered as a well-ordered set—a question which Cantor thinks must be answered in the affirmative. It appears to me most desirable to obtain a direct proof of this remarkable statement of Cantor's, perhaps by actually giving an arrangement of numbers such that in every partial system a first number can be pointed out.

2. The compatibility of the arithmetical axioms

When we are engaged in investigating the foundations of a science, we must set up a system of axioms which contains an exact and complete description of the relations subsisting between the elementary ideas of that science. The axioms so set up are at the same time the definitions of those elementary ideas ; and no statement within the realm of the science whose foundation we are testing is held to be correct unless it can be derived from those axioms by means of a finite number of logical steps. Upon closer consideration the question arises: *Whether, in any way, certain statements of individual axioms depend upon one another, and whether the axioms may not therefore contain certain parts in common, which must be isolated if one wishes to arrive at a system of axioms that shall be altogether independent of one another.*

But above all I wish to designate the following as the most important among the numerous questions which can be asked with regard to the axioms: *To prove that they are not contradictory, that is, that a finite number of logical steps based upon them can never lead to contradictory results.*

In geometry, the proof of the compatibility of the axioms can be effected by constructing a suitable field of numbers, such that analogous relations between the numbers of this field correspond to the geometrical axioms. Any contradiction in the deductions from the geometrical axioms must therefore be recognizable in the arithmetic of this field of numbers. In this way the desired proof of the compatibility of the geometrical axioms is made to depend upon the theorem of the compatibility of the arithmetical axioms.

On the other hand a direct method is needed for the proof of the compatibility of the arithmetical axioms. The axioms of arithmetic are essentially nothing other than the known rules of calculation, with the addition of the axiom of continuity. I recently collected them[3] and in so doing replaced the axiom of continuity by two simpler axioms, namely, the well-known axiom of Archimedes, and a new axiom essentially as follows : that numbers form a system of things which is capable of no further extension, as long as all the other axioms hold (axiom of completeness). I am convinced that it must be possible to find a direct proof for the compatibility of the arithmetical axioms by means of a careful study and suitable modification of the known methods of reasoning in the theory of irrational numbers.

To show the significance of the problem from another point of view, I add the following observation : If contradictory attributes be assigned to a concept, I say, that *mathematically the concept does not exist.* So, for example, a real number whose square is -1 does not exist mathematically. But if it can be proved that the attributes assigned to the concept can never lead to a contradiction by the application of a finite number of logical processes, I say that the mathematical existence of the concept (for example, of a number or a function which satisfies certain conditions) is thereby proved. In the case before us, where we are concerned with the axioms of real numbers in arithmetic, the proof of the compatibility of the axioms is at the same time the proof of the mathematical existence of the complete system of real numbers or of the continuum. Indeed, when the proof for the compatibility of the axioms shall be fully accomplished, the doubts which have been expressed occasionally as to the existence of the complete system of real numbers will become totally groundless. The totality of real numbers, i.e., the continuum according to the point of view just indicated, is not the totality of all possible series in decimal fractions, or of all possible laws according to which the elements of a fundamental sequence may proceed. It is rather a system of things whose mutual relations are governed by the axioms set up and for which all propositions, and only those, are true which can be derived from the axioms by a finite number of logical processes. In my opinion, the concept of the continuum is strictly logically tenable in this sense only. It seems to

---

[3] [Über den Zahlbegriff, JJG] *Jahresbericht der Deutschen Mathematiker-Vereinigung*, vol. 8 (1900), p. 180.

me, indeed, that this corresponds best also to what experience and intuition tell us. The concept of the continuum or even that of the system of all functions exists, then, in exactly the same sense as the system of integral, rational numbers, for example, or as Cantor's higher classes of numbers and cardinal numbers. For I am convinced that the existence of the latter, just as that of the continuum, can be proved in the sense I have described; unlike the system of all cardinal numbers or of all Cantor's alephs, for which, as may be shown, a system of axioms, compatible in my sense, cannot be set up. Either of these systems is, therefore, according to my terminology, mathematically non-existent.

From the field of the foundations of geometry I should like to mention the following problem:

3. The equality of the volumes of two tetrahedra of equal bases and equal altitudes

In two letters to Gerling, Gauss[4] expresses his regret that certain theorems of solid geometry depend upon the method of exhaustion, i.e., in modern phraseology, upon the axiom of continuity (or upon the axiom of Archimedes). Gauss mentions in particular the theorem of Euclid, that triangular pyramids of equal altitudes are to each other as their bases. Now the analogous problem in the plane has been solved.[5] Gerling also succeeded in proving the equality of volume of symmetrical polyhedra by dividing them into congruent parts. Nevertheless, it seems to me probable that a general proof of this kind for the theorem of Euclid just mentioned is impossible, and it should be our task to give a rigorous proof of its impossibility. This would be obtained, as soon as we succeeded *in specifying two tetrahedra of equal bases and equal altitudes which can in no way be split up into congruent tetrahedra, and which cannot be combined with congruent tetrahedra to form two polyhedra which themselves could be split up into congruent tetrahedra.*[6]

---

[4] *Werke*, vol. 8, pp. 241 and 244.

[5] Cf., beside earlier literature, Hilbert, *Grundlagen der Geometrie*, Leipzig, 1899, ch. 4. [Translation by Townsend, Chicago, 1902, MWN.]

[6] Since this was written Herr Dehn has succeeded in proving this impossibility. See his note: 'Ueber raumgleiche Polyeder,' in *Nachrichten d. K. Gesellsch. d. Wiss. zu Göttingen*, 1900, and a paper soon to appear in the *Mathematische Annalen* [vol. 55, pp. 465–478, MWN].

4. Problem of thee straight line as the shortest distance between two points

Another problem relating to the foundations of geometry is this: If from among the axioms necessary to establish ordinary Euclidean geometry, we exclude the axiom of parallels, or assume it as not satisfied, but retain all other axioms, we obtain, as is well known, the geometry of Lobachevsky (hyperbolic geometry). We may therefore say that this is a geometry standing next to Euclidean geometry. If we require further that that axiom be not satisfied whereby, of three points of a straight line, one and only one lies between the other two, we obtain Riemann's (elliptic) geometry, so that this geometry appears to be the next after Lobachevsky's. If we wish to carry out a similar investigation with respect to the axiom of Archimedes, we must look upon this as not satisfied, and we arrive thereby at the non-Archimedean geometries which have been investigated by Veronese and myself. The more general question now arises: Whether from other suggestive standpoints geometries may not be devised which, with equal right, stand next to Euclidean geometry. Here I should like to direct your attention to a theorem which has, indeed, been employed by many authors as a definition of a straight line, viz., that the straight line is the shortest distance between two points. The essential content of this statement reduces to the theorem of Euclid that in a triangle the sum of two sides is always greater than the third side — a theorem which, as is easily seen, deals solely with elementary concepts, i.e., with such as are derived directly from the axioms, and is therefore more accessible to logical investigation. Euclid proved this theorem, with the help of the theorem of the exterior angle, on the basis of the congruence theorems. Now it is readily shown that this theorem of Euclid cannot be proved solely on the basis of those congruence theorems which relate to the application of segments and angles, but that one of the theorems on the congruence of triangles is necessary. We are asking, then, for a geometry in which all the axioms of ordinary Euclidean geometry hold, and in particular all the congruence axioms except the one of the congruence of triangles (or all except the theorem of the equality of the base angles in the isosceles triangle), and in which, besides, the proposition that in every triangle the sum of two sides is greater than the third is assumed as a particular axiom.

One finds that such a geometry really exists and is no other than that which Minkowski constructed in his book, *Geometrie der Zahlen*,[7] and made the basis of his arithmetical investigations. Minkowski's is therefore also a geometry standing next to the ordinary Euclidean geometry; it is essentially characterized by the following stipulations :

1. The points which are at equal distances from a fixed point O lie on a convex closed surface of the ordinary Euclidean space with O as a center.

2. Two segments are said to be equal when one can be carried into the other by a translation of the ordinary Euclidean space.

In Minkowski's geometry the axiom of parallels also holds. By studying the theorem of the straight line as the shortest distance between two points, I arrived[8] at a geometry in which the parallel axiom does not hold, while all other axioms of Minkowski's geometry are satisfied. The theorem of the straight line as the shortest distance between two points and the essentially equivalent theorem of Euclid about the sides of a triangle, play an important part not only in number theory but also in the theory of surfaces and in the calculus of variations. For this reason, and because I believe that the thorough investigation of the conditions for the validity of this theorem will throw a new light upon the idea of distance, as well as upon other elementary ideas, e. g., upon the idea of the plane, and the possibility of its definition by means of the idea of the straight line, *the construction and systematic treatment of the geometries here possible seem to me desirable.*

5. Lie's concept of continuous group of transformations without the assumption of the differentiability of the functions defining the group

It is well known that Lie, with the aid of the concept of continuous groups of transformations, has set up a system of geometrical axioms and, from the standpoint of his theory of groups, has proved that this system of axioms suffices for geometry. But since Lie assumes, in the very foundation of his theory, that the functions defining his group can be differentiated, it remains undecided in Lie's development,

---

[7]   Leipzig, 1896.

[8]   [Hilbert, (1894b) JJG] *Mathematische Annalen*, 46, p. 91.

whether the assumption of the differentiability in connection with the question as to the axioms of geometry is actually unavoidable, or whether it may not appear rather as a consequence of the group concept and the other geometrical axioms. This consideration, as well as certain other problems in connection with the arithmetical axioms, brings before us the more general question : How far is Lie's concept of continuous groups of transformations approachable in our investigations without the assumption of the differentiability of the functions.

Lie defines a finite continuous group of transformations as a system of transformations

$$x_i' = f_i(x_1, \ldots, x_n ; a_1, \ldots a_r) \qquad (i = 1, \ldots, n)$$

having the property that any two arbitrarily chosen transformations of the system, as

$$x_i' = f_i(x_1, \ldots, x_n ; a_1, \ldots a_r)$$
$$x_i'' = f_i(x_1', \ldots, x_n' ; b_1, \ldots b_r)$$

applied successively result in a transformation which also belongs to the system, and which is therefore expressible in the form

$$x_i'' = f_i\{f_1(x,a), \ldots, f_n(x,a); b_1, \ldots, b_r\} = f_i(x_1, \ldots, x_n ; c_1, \ldots, c_r),$$

where $c_1, \ldots, c_r$ are certain functions of $a_1, \ldots, a_r$ and $b_1, \ldots, b_r$.

The group property thus finds its full expression in a system of functional equations and of itself imposes no additional restrictions upon the functions $f_1, \ldots, f_n; c_1, \ldots, c_r$. Yet Lie's further treatment of these functional equations, viz., the derivation of the well-known fundamental differential equations, assumes necessarily the continuity and differentiability of the functions defining the group.

As regards continuity: this postulate will certainly be retained for the present—if only with a view to the geometrical and arithmetical applications, in which the continuity of the functions in question appears as a consequence of the axiom of continuity. On the other hand the differentiability of the functions defining the group contains a postulate which, in the geometrical axioms, can be expressed only in a rather forced and complicated manner. Hence there arises the question whether, through the introduction of suitable new variables and parameters, the group can always be transformed into one whose defining functions are differentiable; or whether, at least, with the help of certain simple assumptions, a transformation is possible into groups admitting Lie's methods. A reduction to analytic groups is, according to a theorem announced by Lie[9] but first proved by Schur[10], always possible when the group is transitive and the existence of the first and certain second derivatives of the functions defining the group is assumed.

For infinite groups the investigation of the corresponding question is, I believe, also of interest. Moreover we are thus led to the wide and interesting field of functional equations which have been heretofore investigated usually only under the assumption of the differentiability of the functions involved. In particular the functional equations treated by Abel[11] with so much ingenuity, the difference equations, and other equations occurring in the literature of mathematics, do not directly involve anything which necessitates the requirement of the differentiability of the accompanying functions. In the search for certain existence proofs in the calculus of variations I came directly upon the problem: to prove the differentiability of the function under consideration from the existence of a difference equation. In all these cases, then, the problem arises : *In how far are the assertions which we can make in the case of differentiable functions true under proper modifications without this assumption?*

It may be further remarked that H. Minkowski in his above-mentioned *Geometry of Numbers* starts with the functional equation

$$f(x_1 + y_1, \ldots, x_n + y_n) \leq f(x_1, \ldots, x_n) + f(y_1, \ldots, y_n)$$

and from this actually succeeds in proving the existence of certain differential quotients for the function in question.

On the other hand I wish to emphasize the fact that there certainly exist analytical functional equations whose sole solutions are non-differentiable functions. For example a uniform continuous non-differentiable function $\varphi(z)$ can be constructed which represents the only solution of the two functional equations

$$\varphi(x + \alpha) - \varphi(x) = f(x), \qquad \varphi(x + \beta) - \varphi(x) = 0,$$

where α and β are two real numbers, and $f(x)$ denotes, for all the real values of x, a regular analytic uniform function. Such functions are

---

[9] Lie–Engel, *Theorie der Transformationsgruppen*, vol. 3, Leipzig, 1893, §§82, 144.

[10] 'Ueber den analytischen Charakter der eine endliche Kontinuierliche Transformationsgruppen darstellenden Funktionen,' *Math. Annalen*, vol. 41.

[11] *Werke*, vol. l, pp. l, 61, 389.

obtained in the simplest manner by means of trigonometrical series by a process similar to that used by Borel (according to a recent announcement of Picard)[12] for the construction of a doubly periodic, non-analytic solution of a certain analytic partial differential equation.

6. Mathematical treatment of the axioms of physics

The investigations on the foundations of geometry suggest the problem: To treat in the same manner, by means of axioms, those physical sciences in which mathematics plays an important part; in the first rank are the theory of probabilities and mechanics.

As to the axioms of the theory of probabilities,[13] it seems to me desirable that their logical investigation should be accompanied by a rigorous and satisfactory development of the method of mean values in mathematical physics, and in particular in the kinetic theory of gases.

Important investigations by physicists on the foundations of mechanics are at hand; I refer to the writings of Mach,[14] Hertz[15], Boltzmann[16] and Volkmann[17]. It is therefore very desirable that the discussion of the foundations of mechanics be taken up by mathematicians also. Thus Boltzmann's work on the principles of mechanics suggests the problem of developing mathematically the limiting processes, there merely indicated, which lead from the atomistic view to the laws of motion of continua. Conversely one might try to derive the laws of the motion of rigid bodies by a limiting process from a system of axioms depending upon the idea of continuously varying conditions of a material filling all space continuously, these conditions being defined by parameters: For the question as to the equivalence of different systems of axioms is always of great theoretical interest.

If geometry is to serve as a model for the treatment of physical axioms, we shall try first by a small number of axioms to include as

---

[12] 'Quelques théories fondamentales dans l'analyse mathématique,' Conférences faites à Clark University, *Revue générale des Sciences*, 1900, p. 22.

[13] Cf. Bohlmann, 'Ueber Versicherungsmathematik', from the collection : Klein and Riecke, *Ueber angewandte Mathematik and Physik*, Leipzig, 1900.

[14] *Die Mechanik in ihrer Entwickelung*, Leipzig, 4th edition, 1901.

[15] *Die Prinzipien der Mechanik*, Leipzig, 1894.

[16] *Vorlesungen über die Principe der Mechanik*, Leipzig 1897.

[17] *Einführung in das Studium der theoretischen Physik*, Leipzig, 1900.

large a class as possible of physical phenomena, and then by adjoining new axioms to arrive gradually at the more special theories—whereby, perhaps a principle of subdivision can be derived from Lie's profound theory of infinite transformation groups. As he has in geometry, the mathematician will not merely have to take account of those theories coming near to reality, but also of all logically possible theories. He must be always alert to obtain a complete survey of all conclusions derivable from the system of axioms assumed.

Further, the mathematician has the duty to test exactly in each instance whether the new axioms are compatible with the previous ones. The physicist, as his theories develop, often finds himself forced by the results of his experiments to make new hypotheses, while he depends, with respect to the compatibility of the new hypotheses with the old axioms, solely upon these experiments or upon a certain physical intuition, a practice which is not admissible in the rigorously logical building up of a theory. The desired proof of the compatibility of all assumptions seems to me also of importance, because the effort to obtain such proof always forces us most effectively to an exact formulation of the axioms.

So far we have considered only questions concerning the foundations of the mathematical sciences. Indeed, the study of the foundations of a science is always particularly attractive, and the testing of these foundations will always be among the foremost problems of the investigator. Weierstrass once said 'The final object always to be kept in mind is to arrive at a correct understanding of the foundations of the science . . . But to make any progress in the sciences the study of particular problems is, of course indispensable.' In fact, a thorough understanding of its special theories is necessary for the successful treatment of the foundations of the science. Only that architect is in the position to lay a sure foundation for a structure who knows its purpose thoroughly and in detail. So we turn now to the special problems of the separate branches of mathematics and consider first arithmetic and algebra.

7. Irrationality and transcendence of certain numbers

Hermite's arithmetical theorems on the exponential function and their extension by Lindemann are certain of the admiration of all generations of mathematicians. But the task at once presents itself of penetrating further along the path here entered, as A. Hurwitz has already done in two interesting papers[18], 'Ueber arithmetische Eigenschaften gewisser transzendenter Funktionen.' I should like, therefore, to sketch a class of problems which, in my opinion, should be taken up here next. That certain special transcendental functions, important in analysis, take algebraic values for certain algebraic arguments, seems to us particularly remarkable and worthy of thorough investigation. Indeed, we expect transcendental functions to assume, in general, transcendental values for even algebraic arguments; and, although it is well known that there exist integral transcendental functions which even have rational values for all algebraic arguments, we shall still consider it highly probable that the exponential function $e^{i\pi z}$, for example, which evidently has algebraic values for all rational arguments z, will on the other hand always take transcendental values for irrational algebraic values of the argument z. We can also give this statement a geometrical form, as follows:

If, in an isosceles triangle, the ratio of the base angle to the angle at the vertex be algebraic but not rational, then the ratio between base and side is always transcendental.

In spite of the simplicity of this statement and of its similarity to the problems solved by Hermite and Lindemann, I consider the proof of this theorem very difficult ; as also the proof that

The expression α^{β}, for an algebraic base (and an irrational algebraic exponent β, e.g., the number $2^{\sqrt{i}}$ or $e^{\pi} = i^{-2i}$, always represents a transcendental or at least an irrational number.

It is certain that the solution of these and similar problems must lead us to entirely new methods and to a new insight into the nature of special irrational and transcendental numbers.

8. Problems of prime numbers

Essential progress in the theory of the distribution of prime numbers has lately been made by Hadamard, de la Vallée-Poussin, Von Mangoldt and others. For the complete solution, however, of the problems set up by Riemann's paper 'Ueber die Anzahl der Primzahlen unter einer gegebenen Grösse', it still remains to prove the

---

[18] *Math. Annalen*, vols. 22, 32 (1883, 1888).

correctness of an exceedingly important statement of Riemann, viz., *that the zeros of the function, $\zeta(s)$, defined by the series*

$$\zeta(s) = 1 + \frac{1}{2^s} + \frac{1}{3^s} + \frac{1}{4^s} + \cdots$$

all have the real part ½, except for the well-known negative integral real zeros. As soon as this proof has been successfully established, the next problem would consist in testing more exactly Riemann's infinite series for the number of primes below a given number and, especially, *to decide whether the difference between the number of primes below a number z and the integral logarithm of z does in fact become infinite of an order not greater than* ½ *in x.*[19] Further, we should determine whether the occasional condensation of prime numbers which has been noticed in counting primes is really due to those terms of Riemann's formula which depend upon the first complex zeros of the function $\zeta(s)$.

After an exhaustive discussion of Riemann's prime number formula, perhaps we may sometime be in a position to attempt the rigorous solution of Goldbach's problem,[20] viz., whether every integer is expressible as the sum of two positive prime numbers; and further to attack the well-known question, whether there are an infinite number of pairs of prime numbers with the difference 2, or even the more general problem, whether the linear Diophantine equation

$$ax + by + c = 0$$

(with given integral coefficients each prime to the others) is always solvable in prime numbers x and y.

But the following problem seems to me of no less interest and perhaps of still wider range: *To apply the results obtained for the distribution of rational prime numbers to the theory of the distribution of ideal primes in a given number-field k*—a problem which looks toward the study of the function $\zeta_k(s)$ belonging to the number-field and defined by the series

$$\zeta_k(s) = \sum \frac{1}{n(j)^s},$$

where the sum extends over all ideals j of the given realm k, and $n(j)$ denotes the norm of the ideal j.

---

[19] Cf. an article by H. von Koch, which is soon to appear in *Mathematische Annalen* [Vol. 55, p. 441, MWN].

[20] Cf. P. Stäckel: 'Über Goldbach's empirisches Theorem,' *Nachrichten d. K. Ges. d. Wiss. zu Göttingen*, 1896, and Landau, ibid., 1900.

I may mention three more special problems in number theory : one on the laws of reciprocity. one on Diophantine equations, and a third from the realm of quadratic forms.

9. Proof of the most general law of reciprocity in any number field

For any field of numbers the law of reciprocity is to be proved for the residues of the ℓ th power, when ℓ denotes an odd prime, and further when ℓ is a power of 2 or a power of an odd prime.

The law, as well as the means essential to its proof, will, I believe, result from suitably generalizing the theory of the field of the ℓ th roots of unity,[21] developed by me, and my theory of relative quadratic fields.[22]

10. Determination of the solvability of Diophantine equation

Given a Diophantine equation with any number of unknown quantities and with rational integral numerical coefficients: *To devise a process according to which it can be determined in a finite number of operations whether the equation is solvable in rational integers.*

11. Quadratic forms with any algebraic numerical coefficients

Our present knowledge of the theory of quadratic number fields[23] puts us in a position to attack successfully the theory of quadratic forms with any number of variables and with any algebraic numerical coefficients. This leads in particular to the interesting problem: to solve a

---

[21] *Jahresber. d. Deutschen Math.-Vereinigung*, 'Ueber die Theorie der algebraischen Zahlkörper,' vol. 4 (1897), Part V.

[22] *Mathematische Annalen*, vol. 51 and *Nachrichten d. K. Ges. d. Wiss. zu Göttingen*, 1898.

[23] Hilbert, 'Ueber den Dirichlet'schen biquadratischen Zahlenkörper,' *Math. Annalen*, vol. 45; 'Ueber die Theorie der relativquadratischen Zahlkörper,' *Jahresber. d. Deutschen Mathematiker-Vereinigung*, 1897, and *Math. Annalen*, vol. 51; 'Ueber die Theorie der relativ-Abelschen Körper,' *Nachrichten d. K. Ges. d. Wiss. zu Göttingen*, 1898 ; Grundlagen der Geometrie, Leipzig, 1899, Chap. VIII, § 83 [Translation by Townsend, Chicago, 1902 MWN]. Cf. also the dissertation of G. Rückle, Göttingen, 1901.

given quadratic equation with algebraic numerical coefficients in any number of variables by integral or fractional numbers belonging to the algebraic realm of rationality determined by the coefficients.

The following important problem may form a transition to algebra and the theory of functions:

12. Extension of Kronecker's theorem on abelian fields to any algebraic domain of rationality

The theorem that every abelian number field arises from the domain of rational numbers by the composition of fields of roots of unity is due to Kronecker. This fundamental theorem in the theory of integral equations contains two statements, namely:

First. It answers the question as to the number and existence of those equations which have a given degree, a given abelian group and a given discriminant with respect to the domain of rational numbers.

Second. It states that the roots of such equations form a domain of algebraic numbers which coincides with the domain obtained by assigning to the argument z in the exponential function $e^{i\pi z}$ all rational numerical values in succession.

The first statement is concerned with the question of the determination of certain algebraic numbers by their groups and their branching. This question corresponds, therefore, to the known problem of the determination of algebraic functions corresponding to given Riemann surfaces. The second statement furnishes the required numbers by transcendental means, namely, by the exponential function $e^{i\pi z}$.

Since the domain of the imaginary quadratic number fields is the simplest after the domain of rational numbers, the problem arises of extending Kronecker's theorem to this case. Kronecker himself has made the assertion that the abelian equations in the domain of a quadratic field are given by the transformation equations of elliptic functions with singular moduli, so that the elliptic function assumes here the same rôle as the exponential function in the former case. The proof of Kronecker's conjecture has not yet been furnished; but I believe that it must be obtainable without very great difficulty on the basis of the theory of complex multiplication developed by H. Weber[24] with the

---

[24] *Elliptische Functionen und algebraische Zahlen.* Braunschweig, 1891.

help of the purely arithmetical theorems on class fields which I have established.

Finally, the extension of Kronecker's theorem to the case when, *in place of the domain of rational numbers or of the imaginary quadratic field, any algebraic field whatever is laid down as domain of rationality,* seems to me of the greatest importance. I regard this problem as one of the most profound and far-reaching in the theory of numbers and of functions.

The problem is found to be accessible from many standpoints. I regard as the most important key to the arithmetical part of this problem the general law of reciprocity for residues of ℓ th powers within any given number field.

As to the function-theoretical part of the problem, the investigator in this attractive region will be guided by the remarkable analogies which are noticeable between the theory of algebraic functions of one variable and the theory of algebraic numbers. Hensel[25] has proposed and investigated the analogue in the theory of algebraic numbers to the development in power series of an algebraic function; and Landsberg[26] has treated the analogue of the Riemann–Roch theorem. The analogy between the genus of a Riemann surface and that of the class number of a field of numbers is also evident. Consider a Riemann surface of genus $p = 1$ (to touch on the simplest case only) and on the other hand a number field of class $h = 2$. To the proof of the existence of an integral everywhere finite on the Riemann surface, corresponds the proof of the existence of an integer α in the number $\sqrt{\alpha}$ field such that the number represents a quadratic field, relatively unbranched with respect to the fundamental field. In the theory of algebraic functions, the method of boundary values serves, as is well known for the proof of Riemann's existence theorem. In the theory of number fields also, the proof of the existence of just this number α offers the greatest difficulty. This proof succeeds with indispensable assistance from the theorem that in the number field there are always prime ideals with given residue characters. This latter fact is therefore the number-theoretic analogue of the problem of boundary values.

---

[25] *Jahresber. d. Deutschen Math.-Vereinigung,* vol. 6, and an article soon to appear in the *Mathematische Annalen* [Vol. 55 p. 301] : 'Ueber die Entwickelung der algebraischen Zahlen in Potenzreihen.'

[26] *Mathematische Annalen,* vol. 50 (1898).

The equation of Abel's theorem in the theory of algebraic functions expresses, as is well known, the necessary and sufficient condition that the points in question on the Riemann surface are the zeros of an algebraic function belonging to the surface. The exact analogue of Abel's theorem, in the theory of the number field of class $h = 2$, is the equation of the law of quadratic reciprocity[27]

$$\left(\frac{\alpha}{j} \right) = + 1$$

which declares that the ideal j is a principal ideal of the number field when and only when the quadratic residue of the number α with respect to the ideal j is positive.

It will be seen that in the problem just sketched the three fundamental disciplines of mathematics, number theory, algebra and function theory, come into closest touch with one another, and I am certain that the theory of analytical functions of several variables in particular would be notably enriched if one should *succeed in finding and discussing those functions which play the part for any algebraic number field corresponding to that of the exponential function in the field of rational numbers and of the elliptic modular functions in the imaginary quadratic number field.*

Passing to algebra, I shall mention a problem from the theory of equations and one to which the theory of algebraic invariants has led me.

13. Impossibility of the solution of the general equation of the 7th degree by means of functions of only two arguments

Nomography[28] deals with the problem of solving equations by means of drawings of families of curves depending on an arbitrary parameter. It is seen at once that every root of an equation whose coefficients depend upon only two parameters, that is, every function of two independent variables, can be represented in many ways according to the principle lying at the foundation of nomography. Further, a large class of functions of three or more variables can evidently be represented by

---

[27] Cf. Hilbert, ' Ueber die Theorie der relativ-Abelschen Zahlkörper,' Gött. Nachrichten, 1898.

[28] d'Ocagne, *Traité de Nomographie*, Paris,1899.

this principle alone without the use of variable elements, namely all those which can be generated by forming first a function of two arguments, then equating each of these arguments to a function of two arguments, next replacing each of those arguments in their turn by a function of two arguments, and so on, regarding as admissible any finite number of insertions of functions of two arguments. So, for example, every rational function of any number of arguments belongs to this class of functions constructed by nomographic tables; for it can be generated by the processes of addition, subtraction, multiplication and division and each of these processes produces a function of only two arguments. One sees easily that the roots of all equations which are solvable by radicals in the natural domain of rationality belong to this class of functions; for here the extraction of roots is adjoined to the four arithmetical operations and this, indeed, represents a function of only one argument. Likewise the general equations of the 5th and 6th degrees are solvable by suitable nomographic tables; for, by means of Tschirnhausen transformations, which require only extraction of roots, they can be reduced to a form where the coefficients depend upon two parameters only.

Now it is probable that the root of the equation of the seventh degree is a function of its coefficients which does not belong to this class of functions capable of nomographic construction, i.e., that it cannot be constructed by a finite number of insertions of functions of two arguments. In order to prove this, the proof would be necessary *that the equation of the seventh degree $f^7 + xf^3 + yf^2 + zf + 1 = 0$ is not solvable with the help of any continuous functions of only two arguments.* I may be allowed to add that I have satisfied myself by a rigorous process that there exist analytical functions of three arguments x, y, z which cannot be obtained by a finite chain of functions of only two arguments.

By employing auxiliary movable elements, nomography succeeds in constructing functions of more than two arguments, as d'Ocagne has recently proved in the case of the equation of the 7th degree.[29]

---

[29] 'Sur la résolution nomographique de l'équation du septième degré.' *Comptes rendus*, Paris, 1900.

14. Proof of the fitness of certain complete systems of functions

In the theory of algebraic invariants, questions as to the finiteness of complete systems of forms deserve, as it seems to me, particular interest. L. Maurer[30] has lately succeeded in extending the theorems on finiteness in invariant theory proved by P. Gordan and myself, to the case where, instead of the general projective group, any subgroup is chosen as the basis for the definition of invariants.

An important step in this direction had been taken already by A. Hurwitz[31] who, by an ingenious process, succeeded in effecting the proof, in its entire generality, of the finiteness of the system of orthogonal invariants of an arbitrary ground form.

The study of the question as to the finiteness of invariants has led me to a simple problem which includes that question as a particular case and whose solution probably requires a decidedly more minutely detailed study of the theory of elimination and of Kronecker's algebraic modular systems than has yet been made.

Let a number m of integral rational functions $X_1, X_2, \ldots, X_m$ of the n variables $x_1, x_2, \ldots, x_n$ be given,

$$X_1 = f_1(x_1, x_2, \ldots, x_n),$$
$$X_2 = f_2(x_1, x_2, \ldots, x_n),$$
$$\ldots\ldots\ldots\ldots\ldots\ldots\ldots$$
$$X_m = f_m(x_1, x_2, \ldots, x_n).$$

Every rational integral combination of $X_1, X_2, \ldots, X_m$ must evidently always become, after substitution of the above expressions, a rational integral function of $x_1, x_2, \ldots, x_n$. Nevertheless, there may well be rational fractional functions of $X_1, X_2, \ldots, X_m$ which, by the operation of the substitution S become integral functions in $x_1, x_2, \ldots, x_n$. Every such rational function of $X_1, X_2, \ldots, X_m$, which becomes integral in x_1, $x_2, \ldots, x_n$ after the application of the substitution S, I propose to call a relatively integral function of $X_1, X_2, \ldots, X_m$. Every integral function of $X_1, X_2, \ldots, X_m$ is evidently also relatively integral ; further, the sum,

---

[30] Cf. *Sitzungsber. d. K. Acad. d. Wiss. zu München*, 1899, and an article about to appear in the *Mathematische Annalen* [I have not found it, JJG].

[31] 'Ueber die Erzeugung der Invarianten durch Integration,' *Nachrichten d. K. Gesellschaft d. Wiss. zu Göttingen*, 1897.

difference and product of relative integral functions are themselves relatively integral.

The resulting problem is now to decide whether it is always possible *to find a finite system of relatively integral function* $X_1, X_2, \ldots, X_m$ *by which every other relatively integral function of* $X_1, X_2, \ldots, X_m$ *may be expressed rationally and integrally.*

We can formulate the problem still more simply if we introduce the idea of a finite field of integrality. By a finite field of integrality I mean a system of functions from which a finite number of functions can be chosen, in terms of which all other functions of the system are rationally and integrally expressible. Our problem amounts, then, to this: to show that all relatively integral functions of any given domain of rationality always constitute a finite field of integrality.

It naturally occurs to us also to refine the problem by restrictions drawn from number theory, by assuming the coefficients of the given functions $f_1, \ldots, f_m$ to be integers and including among the relatively integral functions of $X_1, X_2, \ldots, X_m$ only such rational functions of these arguments as become, by the application of the substitutions S, rational integral functions of $x_1, x_2, \ldots, x_n$ with rational integral coefficients.

The following is a simple particular case of this refined problem : Let m integral rational functions $X_1, X_2, \ldots, X_m$ of one variable x with integral rational coefficients, and a prime number p be given. Consider the system of those integral rational functions of x which can be expressed in the form

$$\frac{G(X_1, \ldots, X_m)}{p^h}$$

where G is a rational integral function of the arguments $X_1, X_2, \ldots, X_m$ and p^h is any power of the prime number p. Earlier investigations of mine[32] show immediately that all such expressions for a fixed exponent h form a finite domain of integrality. But the question here is whether the same is true for all exponents h, i.e., whether a finite number of such expressions can be chosen by means of which for every exponent h every other expression of that form is integrally and rationally expressible.

From the boundary region between algebra and geometry, I will

[32] *Math. Annalen*, vol. 36 (1890), p. 485.

mention two problems. The one concerns enumerative geometry and the other the topology of algebraic curves and surfaces.

15. Rigorous foundation of Schubert's enumerative calculus

The problem consists in this : *To establish rigorously and with an exact determination of the limits of their validity those geometrical numbers which Schubert[33] especially has determined on the basis of the so-called principle of special position, or conservation of number, by means of the enumerative calculus developed by him.*

Although the algebra of today guarantees, in principle, the possibility of carrying out the processes of elimination, yet for the proof of the theorems of enumerative geometry decidedly more is requisite, namely, the actual carrying out of the process of elimination in the case of equations of special form in such a way that the degree of the final equations and the multiplicity of their solutions may be foreseen.

16. Problem of the topology of algebraic curves and surfaces

The maximum number of closed and separate branches which a plane algebraic curve of the *n*th order can have has been determined by Harnack.[34] There arises the further question as to the relative position of the branches in the plane. As to curves of the 6th order, I have satisfied myself—by a complicated process, it is true—that of the eleven branches which they can have according to Harnack, by no means all can lie external to one another, but that one branch must exist in whose interior one branch and in whose exterior nine branches lie, or inversely. *A thorough investigation of the relative position of the separate branches when their number is the maximum seems to me to be of very great interest, and not less so the corresponding investigation as to the number, form, and position of the sheets of an algebraic surface in space.* Till now, indeed, it is not even known what is the maximum number of sheets which a surface of the 4th order in three dimensional space can really have.[35]

---

[33] *Kalkül der abzählenden Geometrie,* Leipzig,1879.
[34] *Math. Annalen,* vol. 10.
[35] Cf. Rohn, 'Flächen vierter Ordnung,' Preisschriften der Fürstlich Jablonowskischen Gesellschaft, Leipzig, 1886.

In connection with this purely algebraic problem, I wish to bring forward a question which, it seems to me, may be attacked by the same method of continuous variation of coefficients, and whose answer is of corresponding value for the topology of families of curves defined by differential equations. This is the question as to the maximum number and position of Poincaré's limit cycles (*cycles limites*) for a differential equation of the first order and degree of the form

$$\frac{dy}{dx} = \frac{Y}{X}$$

where X and Y are rational integral functions of the nth degree in x and y. Written homogeneously, this is

$$X\left(y\frac{dz}{dt} - z\frac{dy}{dt}\right) + Y\left(z\frac{dx}{dt} - x\frac{dz}{dt}\right) + Z\left(x\frac{dy}{dt} - y\frac{dx}{dt}\right) = 0$$

where X, Y, and Z are rational integral homogeneous functions of the nth degree in x, y, z, and the latter are to be determined as functions of the parameter t.

17. Expression of definite forms by squares

A rational integral function or form in any number of variables with real coefficients such that it becomes negative for no real values of these variables, is said to be definite. The system of all definite forms is invariant with respect to the operations of addition and multiplication, but the quotient of two definite forms—in case it should be an integral function of the variables—is also a definite form. The square of any form is evidently always a definite form. But since, as I have shown,[36] not every definite form can be compounded by addition from squares of forms, the question arises—which I have answered affirmatively for ternary forms[37]—whether every definite form may not be expressed as a quotient of sums of squares of forms. At the same time it is desirable, for certain questions as to the possibility of certain geometrical constructions, to know whether the coefficients of the forms to be used in the expression may always be taken from the realm of rationality given by the coefficients of the form represented.[38]

I mention one more geometrical problem.

[36] *Math. Annalen*, vol. 32.

[37] *Acta Mathematica*, vol. 17.

[38] Cf. Hilbert, *Grundlagen der Geometrie*, Leipzig, 1899, Chap. 7 and in particular § 38.

18. Building up of space from congruent polyhedra

If we enquire for those groups of motions in the plane for which a fundamental region exists, we obtain various answers, according as the plane considered is Riemann's (elliptic), Euclid's, or Lobachevsky's (hyperbolic). In the case of the elliptic plane there is a finite number of essentially different kinds of fundamental regions, and a finite number of congruent regions suffices for a complete covering of the whole plane ; the group consists indeed of a finite number of motions only. In the case of the hyperbolic plane there is an infinite number of essentially different kinds of fundamental regions, namely, the well-known Poincaré polygons. For the complete covering of the plane an infinite number of congruent regions is necessary. The case of Euclid's plane stands between these ; for in this case there is only a finite number of essentially different kinds of groups of motions with fundamental regions, but for a complete covering of the whole plane an infinite number of congruent regions is necessary.

Exactly the corresponding facts are found in space of three dimensions. The fact of the finiteness of the groups of motions in elliptic space is an immediate consequence of a fundamental theorem of C. Jordan,[39] whereby the number of essentially different kinds of finite groups of linear substitutions in n variables does not surpass a certain finite limit dependent upon n. The groups of motions with fundamental regions in hyperbolic space have been investigated by Fricke and Klein in the lectures on the theory of automorphic functions,[40] and finally Fedorov,[41] Schoenflies[42] and lately Rohn[43] have given the proof that there are, in Euclidean space, only a finite number of essentially different kinds of groups of motions with a fundamental region. Now, while the results and methods of proof applicable to elliptic and hyperbolic space hold directly for n-dimensional space also, the generalization of the theorem for Euclidean space seems to offer decided difficulties. The investigation of the following question is therefore desirable: *Is there in n-dimensional Euclidean space also only a finite*

---

[39] *Crelle's Journal*, vol. 84 (1878), and *Atti d. Reale Acad. di Napoli* (1880).

[40] *Leipzig*, 1897. Cf. especially Abschnitt I, Chapters 2 and 3.

[41] *Symmetrie der regelmässigen Systeme von Figuren*, 1890.

[42] *Krystallsysteme und Krystallstruktur*, Leipzig, 1891.

[43] *Mathematische Annalen*, vol. 53.

number of essentially different kinds of groups of motions with a fundamental region?

A fundamental region of each group of motions, together with the congruent regions arising from the group, evidently fills up space completely. The question arises : *Whether polyhedra also exist which do not appear as fundamental regions of groups of motions, by means of which nevertheless by a suitable juxtaposition of congruent copies a complete filling up of all space is possible.* I point out the following question, related to the preceding one, and important to number theory and perhaps sometimes useful to physics and chemistry: How can one arrange most densely in space an infinite number of equal solids of given form, e.g., spheres with given radii or regular tetrahedra with given edges (or in prescribed position), that is, how can one so fit them together that the ratio of the filled to the unfilled space may be as great as possible?

If we look over the development of the theory of functions in the last century, we notice above all the fundamental importance of that class of functions which we now designate as analytic functions—a class of functions which will probably stand permanently in the center of mathematical interest.

There are many different standpoints from which we might choose, out of the totality of all conceivable functions, extensive classes worthy of a particularly thorough investigation. Consider, for example, the class of functions characterized by ordinary or partial algebraic differential equations. It should be observed that this class does not contain the functions that arise in number theory and whose investigation is of the greatest importance. For example, the before-mentioned function $\zeta(s)$ satisfies no algebraic differential equation, as is easily seen with the help of the well-known relation between $\zeta(s)$ and $\zeta(1-s)$, if one refers to the theorem proved by Hölder,[44] that the function $\Gamma(z)$ satisfies no algebraic differential equation. Again, the function of the two variables s and x defined by the infinite series

$$\zeta(s,x) = x + \frac{x^2}{2^s} + \frac{x^3}{3^s} + \frac{x^4}{4^s} + \cdots,$$

---

[44] *Mathematische Annalen*, vol. 28.

which stands in close relation with the function $\zeta(s)$, probably satisfies no algebraic partial differential equation. In the investigation of this question the functional equation

$$x\frac{\partial \zeta(s,x)}{\partial x} = \zeta(s-1,x)$$

will have to be used.

If, on the other hand, we are led by arithmetical or geometrical reasons to consider the class of all those functions which are continuous and indefinitely differentiable, we should be obliged in its investigation to dispense with that pliant instrument, the power series, and with the circumstance that the function is fully determined by the assignment of values in any region, however small. While, therefore, the former limitation of the field of functions was too narrow, the latter seems to me too wide.

The idea of the analytic function on the other hand includes the whole wealth of functions most important to science whether they have their origin in number theory, in the theory of differential equations or of algebraic functional equations, whether they arise in geometry or in mathematical physics; and, therefore, in the entire realm of functions, the analytic function justly holds undisputed supremacy.

19. Are the solutions of regular problems in the calculus of variations always necessarily analytic?

One of the most remarkable facts in the elements of the theory of analytic functions appears to me to be this: That there exist partial differential equations whose integrals are all of necessity analytic functions of the independent variables, that is, in short, equations susceptible of none but analytic solutions. The beat known partial differential equations of this kind are the potential equation

$$\frac{\partial^2 f}{\partial x^2} + \frac{\partial^2 f}{\partial y^2} = 0$$

and certain linear differential equations investigated by Picard;[45] also the equation

$$\frac{\partial^2 f}{\partial x^2} + \frac{\partial^2 f}{\partial y^2} = e^f$$

---

[45] *Journal de l'Ecole Polytechnique*, 1890.

the partial differential equation of minimal surfaces, and others. Most of these partial differential equations have the common characteristic of being the Lagrangian differential equations of certain problems of variation, viz. , of such problems of variation

$$\iint f(p,q,z; x,y) \, dx \, dy = \text{minimum}$$

$$\left[p = \frac{\partial z}{\partial x}, q = \frac{\partial z}{\partial y} \right],$$

as satisfy, for all values of the arguments which fall within the range of discussion, the inequality

$$\frac{\partial^2 F}{\partial p^2} \cdot \frac{\partial^2 F}{\partial q^2} + \left(\frac{\partial^2 F}{\partial p \partial q} \right)^2 > 0,$$

F itself being an analytic function. We shall call this sort of problem a regular variation problem. It is chiefly the regular variation problems that play a rôle in geometry, in mechanics, and in mathematical physics; and the question naturally arises, whether all solutions of regular variation problems must necessarily be analytic functions. In other words, *does every Lagrangian partial differential equation of a regular variation problem have the property of admitting analytic integrals exclusively?* And is this the case even when the function is constrained to assume, as, e. g., in Dirichlet's problem on the potential function, boundary values which are continuous, but not analytic?

I may add that there exist surfaces of constant negative Gaussian curvature which are representable by functions that are continuous and possess indeed all the derivatives, and yet are not analytic ; while on the other hand it is probable that every surface whose Gaussian curvature is constant and positive is necessarily an analytic surface. And we know that the surfaces of positive constant curvature are most closely related to this regular variation problem: To pass through a closed curve in space a surface of minimal area which shall enclose, in connection with a fixed surface through the same closed curve, a volume of given magnitude.

20. The general problem of boundary values

An important problem closely connected with the foregoing is the question concerning the existence of solutions of partial differential equations when the values on the boundary of the region are prescribed. This problem is solved in the main by the keen methods of H. A. Schwarz, C. Neumann, and Poincaré for the differential equation of the potential. These methods, however, seem to be generally not capable of direct extension to the case where along the boundary there are prescribed either the differential coefficients or any relations between these and the values of the function. Nor can they be extended immediately to the case where the inquiry is not for potential surfaces but, say, for surfaces of least area, or surfaces of constant positive Gaussian curvature, which are to pass through a prescribed twisted curve or to stretch over a given ring surface. It is my conviction that it will be possible to prove these existence theorems by means of a general principle whose nature is indicated by Dirichlet's principle. This general principle will then perhaps enable us to approach the question: *Has not every regular variation problem a solution, provided certain assumptions regarding the given boundary conditions are satisfied* (say that the functions concerned in these boundary conditions are continuous and have in sections one or more derivatives), *and provided also if need be that the notion of a solution shall be suitably extended?*[46]

21. Proof of the existence of linear differential equations having a prescribed monodromic group

In the theory of linear differential equations with one independent variable z, I wish to indicate an important problem, one which very likely Riemann himself may have had in mind. This problem is as follows : *To show that there always exists a linear differential equation of the Fuchsian class, with given singular points and monodromic group.* The problem requires the production of n functions of the variable z, regular throughout the complex z plane except at the given singular points; at these points the functions may become infinite of only finite order, and when z describes circuits about these points the functions shall undergo the prescribed linear substitutions. The existence of such differential equations has been shown to be probable by counting the constants, but the rigorous proof has been obtained up to this time only in the particular case where the fundamental equations of the given substitutions have roots all of absolute magnitude unity.

---

[46] Cf. my lecture on Dirichlet's principle in the *Jahresbericht der Deutschen Math.-Vereinigung*, vol. 8 (1900), p. 184.

L. Schlesinger has given this proof,[47] based upon Poincaré's theory of the Fuchsian ζ-functions. The theory of linear differential equations would evidently have a more finished appearance if the problem here sketched could be disposed of by some perfectly general method.

22. Uniformization of analytic relations by means of automorphic functions

As Poincaré was the first to prove, it is always possible to uniformize any algebraic relation between two variables by the use of automorphic functions of one variable. That is, if any algebraic equation in two variables be given, there can always be found for these variables two such single-valued automorphic functions of a single variable that their substitution renders the given algebraic equation an identity. The generalization of this fundamental theorem to any analytic non-algebraic relations whatever between two variables has likewise been attempted with success by Poincaré,[48] though by a way entirely different from that which served him in the special problem first mentioned. From Poincaré's proof of the possibility of uniformising an arbitrary analytic relation between two variables, however, it does not become apparent whether the resolving functions can be determined to meet certain additional conditions. Namely, it is not shown whether the two single valued functions of the one new variable can be so chosen that, while this variable traverses the *regular* domain of those functions, the totality of all regular points of the given analytic field are actually reached and represented. On the contrary it seems to be the case, from Poincaré's investigations, that there are beside the branch points certain others, in general infinitely many other discrete exceptional points of the analytic field, that can be reached only by making the new variable approach certain limiting points of the functions. *In view of the fundamental importance of Poincaré's formulation of the question it seems to me that an elucidation and resolution of this difficulty is extremely desirable.*

In conjunction with this problem comes up the problem of reducing to uniformity an algebraic or any other analytic relation among three

---

[47] *Handbuch der Theorie der linearen Differentialgleichungen*, vol. 2, part 2, No. 366.

[48] *Bull. de la Soc. Math. de France*, vol. 11 (1883).

or more complex variables—a problem which is known to be solvable in many particular cases. Toward the solution of this the recent investigations of Picard on algebraic functions of two variables are to be regarded as welcome and important preliminary studies.

23. Further development of the methods of the calculus of variations

So far, I have generally mentioned problems as definite and special as possible, in the opinion that it is just such definite and special problems that attract us the most and from which the most lasting influence is often exerted upon science. Nevertheless, I should like to close with a general problem, namely with the indication of a branch of mathematics repeatedly mentioned in this lecture—which, in spite of the considerable advance Weierstrass has recently given it, does not receive the general appreciation which, in my opinion, is its due—I mean the calculus of variations.[49]

The lack of interest in this is perhaps due in part to the need of reliable modern text books. So much the more praiseworthy is it then that A. Kneser, in a work published very recently, has treated the calculus of variations from the modern points of view and with regard to the modern demand for rigor.[50]

The calculus of variations is, in the widest sense, the theory of the variation of functions, and as such appears as a necessary extension of the differential and integral calculus. In this sense, Poincaré's investigations of the three body problem, for example, form a chapter in the calculus of variations, in so far as Poincaré derives from known orbits by the principle of variation new orbits of similar character.

I add here a short justification of the general remarks upon the calculus of variations made at the beginning of my lecture.

---

[49] Text-books: Moigno-Lindelöf, *Leçons du calcul des variations*, Paris, 1861, and A. Kneser, *Lehrbuch der Variationsrechnung*, Braunschweig, 1900.

[50] As an indication of the contents of this work, it may here be noted that for the simplest problems Kneser derives sufficient conditions of the extremal even for the case that one limit of integration is variable, and employs the envelope of a family of curves satisfying the differential equations of the problem to prove the necessity of Jacobi's conditions of the extremal. Moreover, it should be noticed that Kneser applies Weierstrass's theory also to the inquiry for the extremal of such quantities as are defined by differential equations.

The simplest problem in the calculus of variations proper is known to consist in finding a function y of a variable x such that the definite integral

$$J = \int_a^b F(y_x, y; x)\,dx, \quad y_x = \frac{dy}{dx}$$

assumes a minimum value compared with the values it takes when y is replaced by other functions of x with the same initial and final values.

The vanishing of the first variation in the usual sense

$$\delta J = 0$$

gives for the desired function y the well-known differential equation

$$\frac{dF_{y_x}}{dx} f_y = 0,$$

$$\left[F_{y_x} = \frac{\partial F}{\partial y_x}, F_y = \frac{\partial F}{\partial y} \right],$$

(1)

In order to investigate more closely the necessary and sufficient criteria for the occurrence of the required minimum, we consider the integral

$$J = \int_a^b \{F + (y_x - p)Fp\}\,dx,$$

$$\left[F = F(p, y; x), F_p = \frac{\partial F(p, y; x)}{\partial p} \right],$$

Now we inquire how p is to be chosen, as a function of x, y, in order that the value of this integral J* shall be independent of the path of integration, i.e. , of the choice of the function y of the variable x.

The integral J^* has the form

$$J = \int_a^b \{Ay_x - B\}\,dx,$$

where A and B do not contain y_x , and the vanishing of the first variation

$$\delta J^* = 0$$

in the sense which the new question requires gives the equation

$$\frac{\partial A}{\partial x} + \frac{\partial B}{\partial y} = 0$$

i.e. , we obtain for the function p of the two variables x, y the partial differential equation of the first order

$$\frac{\partial F_p}{\partial x} + \frac{\partial (pF_p - F)}{\partial y} = 0 \qquad (1^*)$$

The ordinary differential equation of the second order (1) and the partial differential equation (1*) stand in the closest relation to each other. This relation becomes immediately clear to us by the following simple transformation

$$\delta J^* = \int_a^b \{F_y \delta y + F_p \delta p + (\delta y_x - \delta p) F_y + (y_x - p) \delta F_p\} \, dx$$

$$= \int_a^b \{F_y \delta y + \delta y_x F_p + (y_x - p) \delta F_p\} \, dx$$

$$= \delta J + \int_a^b (y_x - p) \, \delta F_p \, dx.$$

We derive from this, namely, the following facts : If we construct any simple family of integral curves of the ordinary differential equation (1) of the second order and then form an ordinary differential equation of the first order

$$y_x = p(x, y) \qquad (2)$$

which also admits these integral curves as solutions, then the function $p(x, y)$ is always an integral of the partial differential equation (1*) of the first order ; and conversely, if $p(x, y)$ denotes any solution of the partial differential equation (1*) of the first order, all the non-singular integrals of the ordinary differential equation (2) of the first order are at the same time integrals of the differential equation (1) of the second order, or in short if $y_x = p(x, y)$ is an integral equation of the first order of the differential equation (1) of the second order, $p(x, y)$ represents an integral of the partial differential equation (1*) and conversely; the integral curves of the ordinary differential equation of the second order are therefore, at the same time, the characteristics of the partial differential equation (1*) of the first order.

In the present case we may find the same result by means of a simple calculation ; for this gives us the differential equations (1) and (1*) in question in the form

$$y_{xx}F_{y,y_x} + y_xF_{y,y} + F_{y,x} - Fy = 0, \tag{1}$$

$$(p_x + pp_y)\,F_{pp} + pF_{py} + F_{px} - F_y = 0, \tag{1*}$$

where the lower indices indicate the partial derivatives with respect to x, y, p, yx. The correctness of the affirmed relation is clear from this.

The close relation derived before and just proved between the ordinary differential equation (1) of the second order and the partial differential equation (1*) of the first order, is, as it seems to me, of fundamental significance for the calculus of variations. For, from the fact that the integral J^* is independent of the path of integration it follows that

$$\int_a^b \{F(p) + (y_x - p)\,F_p(p)\}dx = \int_a^b F(\bar{y}_x)dx \tag{3}$$

if we think of the left hand integral as taken along any path y and the right hand integral along an integral curve of the differential equation

$$\bar{y}_x = p(x,\bar{y}).$$

With the help of equation (3) we arrive at Weierstrass's formula

$$\int_a^b F(y_x)dx - \int_a^b F(\bar{y}_x)dx = \int_a^b E(y_x,p)\,dx, \tag{4}$$

where E designates Weierstrass's expression, depending upon y_x, p, y, x,

$$E(y_x, p) = F(y_x) - F(p) - (y_x - p)\,F_p(p).$$

Since, therefore, the solution depends only on finding an integral $p(x, y)$ which is single valued and continuous in a certain neighborhood of the integral curve $\bar{y}$, which we are considering, the developments just indicated lead immediately—without the introduction of the second variation, but only by the application of the polar process to the differential equation (1)—to the expression of Jacobi's condition and to the answer to the question: How far this condition of Jacobi's in conjunction with Weierstrass's condition $E > 0$ is necessary and sufficient for the occurrence of a minimum.

The developments indicated may be transferred without necessitating further calculation to the case of two or more required functions, and also to the case of a double or a multiple integral. So, for example, in the case of a double integral

$$J = \int F(z_x, z_y, z; x, y)\,d\omega, \quad \left[z_x = \frac{\partial z}{\partial x}, z_y = \frac{\partial z}{\partial y} \right],$$

to be extended over a given region ω, the vanishing of the first variation (to be understood in the usual sense)

$$\delta J = 0$$

gives the well-known differential equation of the second order

$$\frac{\partial F_x}{\partial x} + \frac{\partial F_{x_y}}{\partial y} - F_x = 0$$

$$\left[F_{z_x} = \frac{\partial F}{\partial z}, F_z = \frac{\partial F}{\partial z_y}, F_z = \frac{\partial F}{\partial z} \right],$$

for the required function z of x and y.

On the other hand we consider the integral

$$J^* = \int \{F + (z_x - p)F_p + (z_y - q)F_q\}\,d\omega ,$$

$$\left[F = F(p, q, z; x, y), F_p = \frac{\partial F(p, q, z; x, y)}{\partial p}, F_q = \frac{\partial F(p, q, z; x, y)}{\partial q} \right],$$

and inquire, how p and q are to be taken as functions of x, y and z in order that the value of this integral may be independent of the choice of the surface passing through the given closed twisted curve, i.e., of the choice of the function z of the variables x and y.

The integral J^* has the form

$$J^* = \int \{Az_x + Bz_y - C\}\,d\omega$$

and the vanishing of the first variation

$$\delta J^* = 0,$$

in the sense which the new formulation of the question demands, gives the equation

$$\frac{\partial A}{\partial x} + \frac{\partial B}{\partial y} + \frac{\partial C}{\partial z} = 0,$$

i.e. , we find for the functions p and q of the three variables x, y and z the differential equation of the first order

$$\frac{\partial F_p}{\partial x} + \frac{\partial F_q}{\partial y} + \frac{\partial C(pF_p + qF_q - F)}{\partial x} = 0,$$

If we add to this differential equation the partial differential equation

$$p_y + qp_z = qx + pq_x, \tag{I*}$$

resulting from the equations

$$z_x = p(x, y, z)\,, z_y = q(x, y, z)\,,$$

the partial differential equation (I) for the function z of the two variables x and y and the simultaneous system of the two partial differential equations of the first order (I*) for the two functions p and q of the three variables x, y, and z stand toward one another in a relation exactly analogous to that in which the differential equations (1) and (1*) stood in the case of the simple integral.

It follows from the fact that the integral J^* is independent of the choice of the surface of integration z that

$$\int \{F(p,q) + (z_x - p)F_p(p,q) + (z_y - q)F_q(p,q)\}d\omega = \int F(\bar{z}_x, \bar{z}_y)d\omega,$$

if we think of the right hand integral as taken over an integral surface of the partial differential equations

$$\bar{z}_x = p(x, y, \bar{z}), \bar{z}_y = q(x, y, \bar{z});$$

and with the help of this formula we arrive at once at the formula

$$F(z_x, z_y)d\omega - \int F(\bar{z}_x, \bar{z}_y)d\omega = \int E(z_x, z_y, p, q)d\omega,$$

$$[E(z_x, z_y, p, q) = F(z_x, z_y) - F(p,q) - (z_x - p)F_p(p,q) - (z_y - q)F_q(p,q)]\,, \tag{IV}$$

which plays the same rôle for the variation of double integrals as the previously given formula (4) for simple integrals. With the help of this formula we can now answer the question how far Jacobi's condition in conjunction with Weierstrass's condition $E > 0$ is necessary and sufficient for the occurrence of a minimum.

Connected with these developments is the modified form in which A. Kneser,[51] beginning from other points of view, has presented Weierstrass's theory. While Weierstrass employed those integral curves of equation (1) which pass through a fixed point in order to derive sufficient conditions for the extremal, Kneser on the other hand makes use of any simple family of such curves and constructs for every such family a solution, characteristic for that family, of that partial differential equation which is to be considered as a generalization of the Jacobi–Hamilton equation.

The problems mentioned are merely samples of problems, yet they will suffice to show how rich, how manifold and how extensive the

---

[51] Cf. his above-mentioned textbook, §§ 14, 15, 19 and 20.

mathematical science of to-day is, and the question is urged upon us whether mathematics is doomed to the fate of those other sciences that have split up into separate branches, whose representatives scarcely understand one another and whose connection becomes ever more loose. I do not believe this nor wish it. Mathematical science is in my opinion an indivisible whole, an organism whose vitality is conditional upon the connection of its parts. For with all the variety of mathematical knowledge, we are still clearly conscious of the similarity of the logical devices, the relationship of the ideas in mathematics as a whole and the numerous analogies in its different departments. We also notice that, the farther a mathematical theory is developed, the more harmoniously and uniformly does its construction proceed, and unsuspected relations are disclosed between hitherto separate branches of the science. So it happens that, with the extension of mathematics, its organic character is not lost but only manifests itself the more clearly.

But, we ask, with the extension of mathematical knowledge will it not finally become impossible for the individual investigator to embrace all departments of this knowledge? In answer let me point out how thoroughly it is ingrained in mathematical science that every real advance goes hand in hand with the invention of sharper tools and simpler methods which at the same time assist in understanding earlier theories and cast aside older more complicated developments. It is therefore possible for the individual investigator, when he makes these sharper tools and simpler methods his own, to find his way more easily in the various branches of mathematics than is possible in any other science.

The organic unity of mathematics is inherent in the nature of this science, for mathematics is the foundation of all exact knowledge of natural phenomena. That it may completely fulfil this high mission, may the new century bring it gifted masters and many zealous and enthusiastic disciples.

References to the lecture

Abel, N. H. (1881) *Oeuvres Complètes de Niels Hendrik Abel*, l (2 vols), 2nd ed., edited L. Sylow and S. Lie, Christiania.

Bohlmann, I. N. (1900) Ueber Versicherungsmathematik, in C. F. Klein and Riecke, *Ueber angewandte Mathematik and Physik*, Leipzig.

Boltzmann, L. (1897) *Vorlesungen über die Principe der Mechanik*, Leipzig.

Dehn, M. (1900) Ueber raumgleiche Polyeder, *Nachrichten der Königlichen Gesellschaft der Wissenschaften zu Göttingen*, 345–354.

Dehn, M. (1902) Ueber den Rauminhalt, *Mathematische Annalen*, 55, 465–478.

d'Ocagne, M. (1899) *Traité de Nomographie*, Paris.

d'Ocagne, M. (1900) Sur la résolution nomographique de l'équation du septième degré, *Comptes rendus*, Paris.

Fedorov, E. S. (1890) *Simmetria pravil'nich sistem Figur*, St Petersburg, English translation by D. and K. Harker, *Symmetry of crystals*, American Crystallographic Association Monograph, 7, 1971 .

Fricke, R. and Klein, C. F. (1897) *Vorlesungen über die Theorie der Automorphen Functionen*, Teubner, Leipzig and Berlin.

Gauss, C. F. (1813, 1844) Briefe an Gerling, *Werke*, 8, Teubner, Leipzig, pp. 241 and 244.

Harnack, A. (1878) Ueber die Vieltheiligkeit der ebenen algebraischen Curven, *Mathematische Annalen*, 10, 189–198.

Helmholtz, H. L. E. von (1854) 'Ueber die Wechselwirkung der natükräefte und die darauf bezüglischen neuesten Ermittelungen der Phyisk', Vortag, gehalten in Königsberg.

Hensel, K. (1897) Über eine neue Begründung der Theorie der algebraischen Zahlen, *Jahresbericht der Deutschen Mathematiker-Vereinigung*, 6, 83–88.

Hensel, K. (1902) Ueber die Entwickelung der algebraischen Zahlen in Potenzreihen, *Mathematische Annalen*, 55, 301–336.

Hertz, H. (1894) *Die Prinzipien der Mechanik*, Teubner, Leipzig.

Hilbert, D. (1888) Über die Darstellung definiter Formen als Summe von Formenquadraten, *Mathematische Annalen*, 32, 342–350, in *Gesammelte Abhandlungen*, 2, 154–161.

Hilbert, D. (1890) Über die Theorie der algebraischen Formen, *Mathematische Annalen*, **36**, 473–534, in *Gesammelte Abhandlungen*, 2, 199–257.

Hilbert, D. (1893) Über ternäre definite Formen, *Acta Mathematica*, 17, 169–197, in *Gesammelte Abhandlungen*, 2, 345–366.

Hilbert, D. (1894a) Ueber den Dirichlet'schen biquadratischen Zahlenkörper, *Mathematische Annalen*, 45, 309–340, in *Gesammelte Abhandlungen*, 1, 24–52.

Hilbert, D. (1894b) Über die gerade Linie als kürzeste Verbindung zweier Punkte, *Mathematische Annalen*, 46, 91–96, in *Grundlagen der Geometrie*, Anhang I, 126–132.

Hilbert, D. (1897) Ueber die Theorie der algebraischen Zahlkörper, Part V, *Jahresber. d. Deutschen Math.-Vereinigung*, 4, in *Gesammelte Abhandlungen*, 1, 249–363.

Hilbert, D. (1898a) Ueber die Theorie der relativ-Abelschen Körper, *Jahrsbericht den Deutschen mathematiker Vereinigung*, 6, 88–94, in *Gesammelte Abhandlungen*, 1, 364–369.

Hilbert, D. (1898b) Ueber die Theorie der relativ-Abelschen Körper, *Mathematische Annalen*, 51, 1–127, in *Gesammelte Abhandlungen*, 1, 370–482.

Hilbert, D. (1899) *Grundlagen der Geometrie* (*Festschrift zur Einweihung des Göttinger Gauss–Weber Denkmals*), Leipzig; revised 2nd ed. 1903, numerous subsequent editions.

Hilbert, D. (1900) Über den Zahlbegriff, *Jahresbericht der Deutschen Mathematiker-Vereinigung*, 8, 180–184, in *Grundlagen der Geometrie*, 7th ed., Anhang VI, 241–246.

Hölder, O. (1886) Ueber den Eigenschaft der gammafunction keiner algebraischen Differentialgleichungen zu genügen, *Mathematische Annalen* 28, 1–13.

Hurwitz, A. (1883) Ueber arithmetischer Eigenschaften gewisser tranzcendenter Functionen, *Mathematische Annalen* 22, 211–229.

Hurwitz, A. (1888) Ueber arithmetischer Eigenschaften gewisser tranzcendenter Functionen, II *Mathematische Annalen*, 32, 583–588.

Hurwitz, A. (1897) Ueber die Erzeugung der Invarianten durch Integration, *Nachrichten der Königlichen Gesellschaft der Wissenschaften zu Göttingen*, 71–79.

Jordan, C. (1878) Mémoire sur les équations différentielles linéaires à intégrale algébrique, *Journal für die reine und angewandte Mathematik* (*Crelle's Journal*) **84**, 89–215, in *Oeuvres*, II, 13–140.

Jordan, C. (1880) Sur la détermination des groupes d'ordre fini contenus dans le groupe linéaire, *Atti d. Reale Acad. di Napoli.* 8 no. 11 in *Oeuvres de Camille Jordan*, **2**, 177–218 (4 vols, ed. J. Dieudonné).

Kneser, A. (1900) *Lehrbuch der Variationsrechnung*, Braunschweig.

Koch, H. von (1902) Ueber die Riemann'sche Primzahlfunction, *Mathematische Annalen*, **55**, 441–464.

Landau, E. (1900) Ueber die Zahlentheoretische Funktion $\varphi(n)$ und ihre Beziehung zum Goldbachschen Satz, *Nachrichten der Königlichen Gesellschaft der Wissenschaften zu Göttingen*, 177–186.

Landsberg, G. (1898) Über das Analogon der Riemann–Roch'schen Satzes in der Theorie der algebraischen Zahlen, *Mathematische Annalen*, **50**, 577–582.

Lie, M. S. and **Engel, F.** (1893) *Theorie der Transformationsgruppen*, **3**, Teubner, Leipzig.

Mach, E. (1901) *Die Mechanik in ihrer Entwickelung*, Leipzig, 4th ed. English translation.

Maurer, L. (1899) Ueber die Endlichkeit der Invariantensysteme, *Sitzungsber. de. Königlichen Academie der Wissenschaften zu München*, 147–175.

Minkowski, H. (1896) *Geometrie der Zahlen*, Teubner, Leipzig.

Moigno, A. and **Lindelöf, E.** (1861) *Leçons du calcul des variations*, Paris.

Picard, E. (1890) Détermination des intégrales de certaines équations aux dérivées partielles, *Journal de l'Ecole Polytechnique*, **60**.

Picard, E. (1900) Quelques théories fondamentales dans l'analyse mathématique, Conférences faites à Clark University, *Revue générale des Sciences*, **22**.

Poincaré, H. (1883) Sur un théorème de la théorie générale des fonctions, *Bulletin de la Société Mathématique de France*, **11**, 112–125, in *Oeuvres*, **4**, 57–69.

Rohn, K. (1886) Flächen vierter Ordnung, Preisschriften der Fürstlich Jablonowskischen Gesellschaft, Leipzig.

Rohn, K. (1900) Einige Sätze über Regelmässige Punktgruppen, *Mathematische Annalen*, **53**, 440–449.

Rückle, G. (1901) Quadratische Reziprozitätsgesetze in algebraischen Zahlkörpern, Göttingen dissertation.

Schlesinger, L. (1898) *Handbuch der Theorie der linearen Differentialgleichungen*, **2**, Teubner, Leipzig.

Schoenflies, A. (1891) *Krystallsysteme und Krystallstruktur*, Leipzig.

Schubert, H. (1979) *Kalkül der abzählenden Geometrie*, Teubner, Leipzig,1879, rep. with an Introduction by S. Kleiman, Springer Verlag, New York.

Schur, F. (1891) Ueber den analytischen Charakter der eine endliche Kontinuierliche Transformationsgruppen darstellenden Funktionen, *Mathematische Annalen*, **41**, 509–538.

Stäckel, P. (1896) Über Goldbach's empirisches Theorem, *Nachrichten der Königlichen Gesellschaft der Wissenschaften zu Göttingen*, 292–299.

Volkmann (1900) *Einführung in das Studium der theoretischen Physik*, Leipzig.

Weber, H. (1891) *Elliptische Functionen und algebraische Zahlen*, Braunschweig.

Weil, A. (1984) *Number Theory from Hammurapi to Legendre*, Birkhäusen, Boston.

The Hilbert problems 1900–99[1]

1. The Continuum hypothesis.

Shown to be consistent with the usual axioms for set theory by Gödel in 1939–40, and independent of them by Cohen in 1963–64. See Cohen (1966).

2. The compatibility of the axioms of arithmetic.

Gödel (1931) showed that, as usually formulated, arithmetic is incomplete and undecidable.

3. The definition of Euclidean volume

Solved by Dehn (1902) after a partial solution by Bricard (1896).

4. Straight lines and shortest distance

Too vague.

5. Lie groups.

Solved by Gleason (1952) and Montgomery and Zippin (1952).

6. The axiomatisation of physics

Classical mechanics was axiomatised by Hamel in 1903.

Thermodynamics was axiomatised by Carathéodory (1909).

Special relativity was axiomatised by Robb (1914) and Carathéodory (1924) independently.

Probability theory was axiomatised by Kolmogorov (1930).

Quantum field theory was axiomatised by Wightman in the late 1950s.

7. The transcendence of certain numbers.

Solved by Gelfond (1934) and Th. Schneider independently (1934).

---

[1] See Kantor (1996) for some interesting comments and directions for further reading.

8. The Riemann Hypothesis, its generalisations, Goldbach's conjecture
Still unsolved, although many analogous problems have been solved. Goldbach's conjecture is also still unsolved.

9. A generalised reciprocity law
Discovered by Emil Artin (1923).

10. Diophantine equations
Matijasevich (1970) showed that there is no algorithm for these.

11. Arbitrary quadratic forms.
Solved over the rational numbers by Hasse (1923–24), and over the integers by Siegel in the 1930s.

12. Kronecker's Jugentraum and its generalisation.
Abelian class-field theory was created by Takagi in 1920.

13. Nomography.
Solved in the negative by Kolmogorov and Arnold in 1957.

14. The finiteness of systems of invariants.
Solved in the negative by Nagata (1962).

15. Schubert's enumerative calculus.
A rigorous theory is due to van der Waerden in the late 1930s and, in related contexts, a variety of other workers.

16. The topology of curves; the number of boundary cycles.
On the topology of curves the best results are due to Itenberg and Viro (1996).

On limit cycles of flows given by polynomials the best partial results are due to Ilyashenko and Écalle in the early 1990s.

17. Functions as sums of squares.
Solved for real-closed fields by Artin (1927). The upper bound on the number of forms required is due to Pfister (1967) The negative solution in general is due to DuBois (1967).

18. Congruent polyhedra, fundamental domains, and sphere packings.
That only a finite number of groups tessellate space in dimension n is due to Bieberbach (1910). The fundamental domain problem was solved by Reinhardt in dimension 3 and Heesch in dimension 2. The sphere packing problem remains unsolved.

19. Regularity theorems in partial differential equations, 20. Calculus of variations and boundary conditions, and 23. New methods in the calculus of variations.

Problem 19 was first solved by Bernstein in 1904. These problem have since merged and been comprehensively generalised.

21. The Riemann–Hilbert problem

Solved in the negative by Anosov and Bolibruch (1994).

22. The Uniformisation theorem.

Solved by Koebe (1907) and Poincaré independently (1907).

Logical matters[1]

When Hilbert raised the question of finding axioms for arithmetic in 1900, his attention was fixed on eliminating the construction of the real numbers as sets of rational numbers. He wanted an axiom system that codified the known fundamental properties of the real numbers, and which had a unique model. It is not easy to say if he was successful in that aim, because his language lacks the necessary precision. There are axiom systems that we call categorical: they are satisfied by an essentially unique set of objects. But mathematical logicians have taught us that we must always state of any axiom system what system of logic underlies it, and this sophisticated insight was not available to Hilbert in 1900. If one talks in logical terms about a set and its elements, the logic is first-order; if one talks about properties of all subsets of a set, the logic is second-order. Because the properties of sets can be murky, many people consider first-order logic intuitive and second-order logic sophisticated and no more intuitive than set theory itself.

Gödel's completeness theorem of 1930 states that every consistent set of first-order sentences has a model. It does not follow that it is complete, in the sense that any properly formulated statement in the theory is either provable or refutable. What Gödel had in mind is often called semantic completeness. That is to say that when the formal symbols of the system are given an interpretation within a model, some sentences become true. Gödel said such sentences are satisfiable (with respect to the interpretation), and a sentence is valid if its negation is not satisfiable. What he showed was

---

[1] It is a particular pleasure to thank Matt Frank and Moshe Machover for their help with this appendix.

that any valid sentence can be given a formal proof in a finite number of steps.

Nor does the completeness theorem imply that the model is unique, rather the reverse. The Norwegian logician Thoralf Skolem had shown in 1923, as Gödel was perhaps the first to appreciate, that for first-order theories with at least one infinite model there will always be models of different cardinalities. This means in particular that a consistent first-order theory of the real numbers, which we expect and want to have the familiar uncountable model (the 'real' real numbers) also has a countable model. The same is true of set theory itself—a discomfiting result called Skolem's paradox, and we shall consider it briefly below. On the other hand, it was reasonably well known that a second-order theory of the real numbers is categorical. The historian may note that explicit awareness of the second-order nature of the theory came well after the mathematical argument was established, but that is not important.

Gödel's incompleteness theorem of 1931 says that the first-order theory of arithmetic is incomplete. This means that there is no recursive or computable, consistent set of first-order sentences that delivers all of arithmetic. Because the axioms for number theory consist of those describing the logic plus some to describe number theory, and the logical system is complete, this means that the axioms for number theory must be incomplete as Kleene has pointed out.[2] This is one door that opens the way to non-standard models of arithmetic.[3] The door is there in all of the generally used formalisations of mathematics: Gödel explicitly mentioned the Zermelo–Frankel axioms for set theory, the Gödel–von Neumann system, and the Peano axioms suitably enriched.[4]

We are all therefore faced with a choice: to go on or to go back. We can go on, and accept a second-order theory. This has many intuitively acceptable results, but its foundations are murky. Some regard using second-order logic as equivalent to admitting set

---

[2] Kleene (1988) p. 57.
[3] See Kleene (1952).
[4] See Gödel (1986) p. 181.

theory. We can retreat to a first-order system, which has clear foundations, but many counter-intuitive properties. In particular, we must accept that despite the hopes invested in it at the start of the twentieth century by the likes of Hilbert and Weyl, the existence of non-standard models means that 'axiomatic set theory is unable to characterize some of the most basic notions of mathematics, including intuitive set-theoretic notions—except in a merely verbal sense. If mathematics—and in particular the arithmetic of natural numbers—is more than mere verbal discourse, then its reduction to axiomatic set theory somehow fails to do it full justice.'[5]

There is one curious feature of first-order theories that is worth mentioning, and which is due to Tarski. Alfred Tarski was a Polish logician, part of a remarkable flourishing of that subject in newly independent Poland that lasted until the Nazi invasion. In 1930 he was investigating first-order systems of geometry. The axioms speak of points in a geometry, but not of sets of points (such talk is second-order). Hilbert's axioms for geometry were nearly all first-order, but Dedekind's continuity axiom is not. Tarski dropped it. The result is that there are all sorts of models for the geometry—if there are any at all—so it is hard to say what the geometry is about. It is about many things indifferently. But the good news is that the theory is deductively complete: for every statement one can make in the theory, either it or its negation is provable. The provable ones are the ones which are true when the theory is given a coordinate model in which the coordinates are the real numbers. Better yet, there is an algorithm for deciding if any given statement is provable or not, so the theory is decidable in Hilbert's sense.[6]

Tarski also showed that ordinary Euclid geometry, suitably formalised, is neither complete nor decidable. The first-order fragment of it which is decidable, and for which Tarski gave a complete set of axioms, is unable to formulate the concept of a general configuration, and some basic ideas lie beyond it (such as the concept

---

[5] The concluding words of Machover (1996), p. 282.

[6] This works leads directly to the theory of real-closed fields, mentioned on p. NNN.

of area). This forces it into some unnatural formulations of many results.[7] It is indicative of the more elementary and fundamental nature of projective geometry that essentially all theorems of projective geometry, such as Desargues's and Pascal's theorems, can be expressed in Tarski's system. More interestingly for physicists, the English mathematician A.A. Robb had shown how to describe Einstein's theory of special relativity (sometimes called Minkowski space–time) axiomatically in a book in 1914. His axioms make it possible to show that Minkowski space–time geometry can be given a complete set of axioms. Presumably one's emotional response to this result depends on one's inner need for security and one's prior belief that one lives in space and time, or in space–time.

---

[7] Tarski (1959).

Glossary

Algebraic function. An algebraic function is a polynomial expression that defines a variable y as a function of a variable x. A good example is given by the equation for the circle, $x^2 + y^2 - 1 = 0$, which defines y as an algebraic function of x. For each (complex) value of x there are generally two (complex) values of y (which makes y something other than a function in modern terms—historical respect permits the concept to be stretched). In this case they can be found explicitly: $y = \sqrt{1-x^2}$, which makes it clear that there are only two values unless $x = 1$ or -1, when there is only one (the value 0).

Axiom systems. Axiom systems are rules for a collection of objects. They have to be self-consistent if they describe anything at all. There may be essentially only one sort of object that obeys the rules, in which case the axiom system is said to be categorical, or there may be many (there are infinitely many different groups, for example). It may be that a given set of rules is such that any statement that makes sense can either be proved or disproved; such an axiom system is said to be deductively complete.

Axioms. In mathematics axioms specify the rules according to which mathematical objects are manipulated. From a knowledge of the addition and multiplication tables for small integers we all generate the rules for handling any integers, and these can be expressed as axioms (for example, the distributive law: $a(b+c) = ab+ac$). Like the axioms in Euclid's *Elements*, these axioms codify the properties of familiar objects, and license them in novel situations.

Countable and **uncountable** sets: see Box 1.3.

Differential equation. A differential equation is an equation relating a function and some of its derivatives. For example, the function $y = \sin x$ can be differentiated, to give $y' = \cos x$, where

y' denoted the first derivative. This can be differentiated in its turn, to give $y'' = -\sin x$, from which we deduce the differential equation $y'' = -y$. This is a *second-order* differential equation because the second derivative of the function y appears. In a physical context, if y denotes position and x denotes time, then the first derivative suggests velocity and the second derivative acceleration, which is why many significant differential equations are second-order. It is an *ordinary* differential equation because the function depends on a single variable. A differential equation obtained for a function that depends on more than one variable is called a *partial* differential equation. Most differential equations in physics are partial differential equations, and partial differential equations are usually much harder to solve.

Function. A function in mathematics is any quantity whose value depends on another quantity (or quantitites). Good examples (in various notations) are $y = \sin x$, $f(x) = x^2$, or, to display a function mathematicians would like to know much more about, $p(n) =$ the nth prime number. The temperature on the Earth's surface is a function of three variables, latitude, longitude, and time.

The allowed or possible input to the function are said to form the *domain* of the function, the range of its values its *codomain*.

Group. A group is a set of objects which can be combined together in pairs to make a third, and which satisfies other axioms (there is an identity element e such that $e.a = a$ for every a in the group, every element a has an inverse a^{-1} such that $a^{-1}.a = e$, and $a.(b.c) = (a.b).c$ for any three elements a, b, and c in the group). Good examples are the set of all symmetries of an object (see Box 1.2) and the integers combined using addition. A fundamental distinction in group theory is between groups (such as the integers) for which it is always true that $ab = ba$, and those (like the symmetry group of an equilateral triangle) where the order matters and it can happen that $ab \neq ba$. The first kind are called *commutative* (or *Abelian*) groups, the second kind *non-commutative* groups. The first kind are much easier to study.

Infinite sets. Sets with infinitely many members posed an increasing problem when mathematicians wanted to apply logic to discussions of sets.

Integral equation. An integral equation is an equation for an unknown function that involves an integral. For example, this equation for the unknown function $x(t)$: $x(t) = f(t) + \int_0^1 K(t,s,x(s))\mathrm{d}s$. The original problem leading to the equation determines the function K, which is called the *kernel* of the equation.

Logic. Initially the common-sense rules of reasoning we all accept, such as: Socrates is a man, all men are mortal, therefore Socrates is mortal. Problems arise when applying these rules to infinite sets. *Intuitionistic logic*, associated with Brouwer, denies that certain claims of the kind 'Either there is an element of such-and-such a kind in this set, or there is not' apply to infinite sets.

Number theory. Number theory is the study of the integers. A major unsolved problem is Goldbach's conjecture: is every even number greater than 2 the sum of two prime numbers? It splits into various branches, depending on the methods used. Analytic number theory makes subtle use of analysis, typically estimating quantities expressed as integrals. Algebraic number theory makes increasing use of the structural properties of algebra (for example, group theory). It is responsible for stretching the concept of an integer: an *algebraic integer* is the root of a polynomial equation of the form $x^n + a_1 x^{n-1} + \cdots + a_n = 0$ with integer coefficients. The familiar integers are algebraic integers (the integer n is the root of the equation $(x - n = 0)$. Expressions such as $\sqrt{2}$ are also algebraic integers ($\sqrt{2}$ satisfies $x^2 - 2 = 0$).

Set. A set is a collection of distinct objects. Paradoxes arise when this naïve definition is put to use, and mathematicians sought ways of defining sets which allowed for rigorous argument. The result is that a *set* is a special kind of a collection, and some collections are not sets. Given a set, the collection of all of its subsets is again a set.

General references

Alexandrov, P. S. ed. (1979) *Die Hilbertsche Probleme*, German ed. Ostwalds Klassiker der exakten Wissenschaften, **252**, Leipzig.

Anosov, D. V. and **Bolibruch, A. A.** (1994) *The Riemann–Hilbert Problem*, Steklov Institute of Mathematics, Vieweg.

Artin, E. (1927) Über die Zerlegung definiter Funktionen in Quadrate, *Abhandlungen der Mathematische Seminar Hamburg*, **5**, 100–115.

Baker, A. (1984) *A Concise Introduction to the Theory of Numbers*, Cambridge University Press, Cambridge.

Barrow-Green, J. E. (1997) *Poincaré and the Three Body Problem*, American and London Mathematical Societies, History of Mathematics **11**, Providence, Rhode Island.

Bashmakova, I. G. (1997) *Diophantus and Diophantine Equations*, updated by Joseph Silverman, translated from the Russian by Abe Shenitzer with the editorial assistance of Hardy Grant, Washington, DC, Mathematical Association of America, Dolciani mathematical expositions **20**.

Bendixson, I. O. (1901) Sur les courbes définies par des équations différentielles, *Acta Mathematica* **24**, 1–88.

Bernstein, S. (1904) Sur la nature analytique des solutions des équations aux dérivées partielles du second ordre, *Mathematische Annalen*, **59**, 20–76.

Bieberbach, L. (1910) Über die Bewegungsgruppen des n-dimensionalen euklidischen Raumes mit einem endlichen Fundamentalbereich, *Nachrichten der Königlichen Gesellschaft der Wissenschaften zu Göttingen*, 75–84.

Bieberbach, L. (1930) Über den Einfluss von Hilberts Pariser Vortrag über 'Mathematische Probleme', *Die Naturwissenschaften*, 1101–1111.

Biermann, K.-R. (1988) *Die Mathematik und ihre Dozenten an der Berliner Universität, 1810–1933*, Akademie Verlag, Berlin.

Birkhoff, G. D. (1913) A theorem on matrices of analytic functions, *Mathematische Annalen* **74**, 122–133, and a correction to it, p. 161.

Blumenthal, O. (1922) David Hilbert, *Die Naturwissenschaften*, 67–72.

Blumenthal, O. (1935) *Lebensgeschichte*, in D. Hilbert, *Gesammelte Abhandlungen*, 3, Berlin, pp. 388–429.

Boltianskii, V. G. (1978) *Hilbert's Third Problem*, translated by R. A. Silverman, Winston and Sons, New York.

Boltzmann, L. (1900) Über die Entwicklung der Methoden der theoretischen Physik in neuerer Zeit, *Jahresbericht der Deutschen Mathematiker-Vereinigung*, **8**, 71–95.

Bombieri, E. (1974) Variational problems and elliptic equations, *Proceedings of the International Congress of Mathematicians*, Vancouver 1974, 53–56.

Born, M. (1922) Hilbert und die Physik, *Die Naturwissenschaften* **10**, 88–93. Rep. in M. Born, *Ausgewählte Abhandlungen*, Göttingen, Vandenhoek & Ruprecht (1963), **2**, 584–598.

Bottazzini, U. (1988) Fondamenti dell'aritmetica e della geometria, *Storia della scienza moderna e contemporanea*, III.I. UTET.

Bourbaki, N. (1971) The Architecture of Mathematics, in Le Lionnais (1971), pp. 23–36.

Brezis, H. and **Browder, F.** (1998) Partial differential equations in the 20th century, *Advances in Mathematics*, **135**, 76–144.

Bricard, R. (1896) Sur une question de Géométrie relatives aux polyèdres, *Nouvelles Annales des Mathématiques* (3) **5**, 331–334.

Brouwer, (1908) Die Theorie der endlichen kontinuerlichen Gruppen, unabhängig von den Axiomen von Lie, *Fourth International Congress of Mathematicians*, Rome, 296–303.

Brouwer, L. E. J. (1909a) Die Theorie der endlichen kontinuerlichen Gruppen, unabhängig von den Axiomen von Lie, *Mathematische Annalen*, **67**, 246–267.

Brouwer, L. E. J. (1909b) Die Theorie der endlichen kontinuerlichen Gruppen, unabhängig von den Axiomen von Lie (Zweite Mitteilung), *Mathematische Annalen* 69, 181–203.

Browder, F. ed. (1976) *Mathematical Developments Arising from Hilbert's Problems*, Symposia in Pure Mathematics, American Mathematical Society, **28**, Providence, Rhode Island.

Carathéodory, C. (1909) Untersuchungen über die Grundlagen der Thermodynamik, *Mathematische Annalen*, **67**, 355–386.

Carathéodory, C. (1924) *Zur Axiomarik der speziellen Relativitätstheorie*, Sitzungsberichte Preussische Akademie der Wissenschaften, Math-Physik Klasse, 12–27.

Cassidy, D. C. (1992) *Uncertainty; The Life and Science of Werner Heisenberg*, Freeman, New York.

Ceicignani, C. (1998) *Ludwig Boltzmann: the Man Who Trusted Atoms*, Oxford University Press, Oxford.

Chong, C. T. and **Leong, Y. K.** (1986) An Interview with Jean-Pierre Serre, *Mathematical Intelligencer*, **8.**4, 8–13.

Church, A. (1966) Paul J. Cohen and the continuum problem, *Proceedings of the International Congress of Mathematicians*, Moscow 1966, 15–20.

Cohen, P. J. (1966) *Set Theory and the Continuum Hypothesis*, Freeman, New York.

Corry, L. (1992) Nicolas Bourbaki and the concept of mathematical structure, *Synthese*, **92**, 315–348.

Corry, L. (1996) *Modern Algebra and the Rise of Mathematical Structures*, Birkhäuser, Boston and Basel.

Corry, L. (1997), David Hilbert and the axiomatization of physics (1894–1905), *Archive for History of Exact Sciences*, **51**, 83–198.

Corry L. (1999) Hilbert and physics (1900–1915), 145–188, in *The Symbolic Universe; Geometry and Physics 1890–1930*, ed. J.J. Gray, Oxford University Press, Oxford.

Cox, D. A. and **Katz, S.** (1999) *Mirror Symmetry and Algebraic Geometry*, Mathematical Surveys and Monographs **68**, American Mathematical Society, Providence, Rhode Island.

Dauben, J. W. (1979) *Georg Cantor. His Mathematics and Philosophy of the Infinite*, Harvard University Press, Cambridge, Mass.

Dawson, J. W. (1991) The reception of Gödel's incompleteness theorem, in *Perspectives on the History of Mathematical Logic*, pp. 84–100, ed. T. Drucker, Birkhäuser, Boston and Basel.

Dehn, M. (1902) Ueber den Rauminhalt, *Mathematische Annalen*, **55**, 465–478.

Demidov, S. (1993) The Moscow School of the Theory of Functions in the 1930s, 35–54 in *Golden Years of Moscow Mathematics*, ed. S. Zdravkovska, P. L. Duren, American and London Mathematical Societies, History of Mathematics, **6**, Providence, Rhode Island.

Diacu, F. and **Holmes, P.** (1996) *Celestial Encounters, The Origins of Chaos and Stability*, Princeton University Press, Princeton.

Dieudonné, J. (1939) Modern axiomatic methods and the foundations of mathematics, reprinted in Le Lionnais (1971), **2**, 251–266.

Dieudonné J. (1971) David Hilbert (1862–1943), reprinted in Le Lionnais (1971) pp. 304–311.

Douglas, J. (1933) The problem of Plateau, *Bulletin of the American Mathematical Society*, **39**, 227–251.

Dubois, D. W. (1967) Note on Artin's solution of Hilbert's 17th problem, *Bulletin of the American Mathematical Society*, **73**, 540–541.

Du Bois Reymond, E., (1884) *Über die Grenzen des Naturerkennens und die sieben Welträthsel*, von Veit & Co. Leipzig.

Edwards, H. M. (1974) *Riemann's Zeta Function*, Academic Press New York, London.

Edwards, H. M. (1975) The background of Kummer's proof of Fermat's Last Theorem for regular primes, *Archive for History of Exact Sciences*, **14**, 219–236.

Edwards, H. M. (1977) *Fermat's Last Theorem*, Springer Verlag, New York.

Edwards, H. M. (1980) The genesis of ideal theory, *Archive for history of exact sciences* **23**, 321–378.

Encyklopadie der Mathematischen Wissenschaften mit Einschluss ihrer Anwendungen. (*Encyclopedia of Mathematical Sciences including their Applications*).

Edwards, H. M. (1990) *Divisor Theory*, Birkhäuser, Boston and Basel.

Epple, M. (1999) *Die Entstehung der Knotentheorie*, Vieweg, Wiesbaden.

Euclid (1956) *The Thirteen Books of Euclid's Elements*, ed. and tr. Sir T.L. Heath, Cambridge University Press, 3 vols, Dover Reprint, New York.

Fölsing, A. (1993) *Albert Einstein* Suhrkamp, Frankfurt am Main, English translation E. Osers, Viking, Penguin New York.

Frege, G. (1976) *Wissenschaftlicher Briefwechsel*, ed. G. Gabriel, Hamburg.

Frege, G. (1980) *Philosophical and Mathematical Correspondence*, abridged from the German edition by Brian McGuiness and translated by Hans Kaal, The University of Chicago Press, Chicago.

Frei, G. ed. (1985) *Der Briefwechsel David Hilbert–Felix Klein (1886–1918)*, Mit Anmerkungen, Vandenhoeck & Ruprecht, Göttingen.

Fuchs, D. B. (1993) On Soviet Mathematics of the 1950s and 1960s, in *Golden Years of Moscow Mathematics*, pp. 213–222, ed. S. Zdravkovska, P. L. Duren, American and London Mathematical Societies, History of Mathematics, **6**, Providence, Rhode Island.

Gårding, L. (1998) *Mathematics and Mathematicians; Mathematics in Sweden before 1950*, American and London Mathematical Societies, History of Mathematics, **13**, Providence, Rhode Island.

Gauss, C. F. (1917–33) *Werke*, **10.1**, Akademie der Wissenschaften in Göttingen, Leipzig and Berlin.

Gelford, A. O. (1934) On Hilbert's seventh problem, *Dokl. Adad Nauk*, SSSR, **2**, 1–6.

Givant, S. (1991) A portrait of Alfred Tarski, *Mathematical Intelligencer* **13**.3, 16–32.

Gleason, A. M. (1952) Groups without small subgroups, *Annals of Mathematics*, **56**, 193–212.

Gödel, K. (1929) On the completeness of the calculus of logic, Doctoral dissertation University of Vienna, in *Collected Works*, **1**, 62–101.

Gödel, K. (1930) Die Vollständigkeit der Axiome des logischen Funktionenkalküls, *Monatshefte für Mathematik und Physik* **37**, 349–360, translated with a reprint of the original as The completeness of the axioms of the functional calculus, in *Collected Works*, **1**, 102–123.

Gödel, K. (1931) Über formal unentscheidbare Sätze der *Principia mathematica* und verwandter Systeme I, translated with a reprint of the original as On formally undecidable propositions of *Principia mathematica* and related systems I, in *Collected Works*, **1**, 144–195.

Gödel, K. (1947) What is Cantor's continuum problem?, *American Mathematical Monthly* **54**, 515–525 in *Collected Works*, **2**, 176–187.

Gödel, K. (1986) *Collected Works*, 1, *Publications 1929–1936*, numerous editors, Oxford University Press, Oxford.

Gödel, K. (1990) *Collected Works*, 2, *Publications 1938–1974*, numerous editors, Oxford University Press, Oxford.

Gray, J. J. (1991) Did Poincaré say 'Set theory is a disease'?, *The Mathematical Intelligencer*, **13**.1, 19–22.

Gray, J. J. (1998/99) Mathematicians as philosophers of mathematics: Part 1, *For the learning of mathematics*, **18**.3, 20–24, Mathematicians as philosophers of mathematics: Part 2, *For the learning of mathematics*, **19**.1, 28–31.

Gray, J. J. (1999) *Linear differential equations and group theory from Riemann to Poincaré*, Birkhäuser, Boston and Basel.

Hardy, G. H. (1940) *A Mathematician's Apology*, many subsequent editions, Cambridge University Press, Cambridge.

Hawkins, T. (1982) Wilhelm Killing and the Structure of Lie Algebras, *Archive for History of Exact Sciences* **26**, 127–192.

Helmholtz, H von, 1870 On the origin and significance of the axioms of geometry, tr. M. F. Lowe, in *Hermann von Helmholtz, Epistemological Writings*, ed. P. Hertz and M. Schlick, Boston Studies in the physics of science, 37, Reidel, Dordrecht and Boston, 1977.

Hilbert, D. (1888) Über die Darstellung definiter Formen als Summe von Formenquadraten, *Mathematische Annalen*, **32**, 342–350, in *Gesammelte Abhandlungen* 2, 154–161.

Hilbert, D. (1891) Über die reellen Züge algebraischer Kurven, *Mathematische Annalen* **38**, 115–138, in *Gesammelte Abhandlungen* 2, 415–436.

Hilbert, D. (1893) Über die Theorie der algebraischen Invarianten, *Mathematical papers read at the international Mathematical Congress, Chicago, 1893*, 116–124, Macmillan, New York, in *Gesammelte Abhandlungen* 2, 276–383.

Hilbert, D. (1897) Die Theorie der algebraischen Zahlkörper (*Zahlbericht*), *Jahrsbericht den Deutschen mathematiker Vereinigung*, **4**, 175–546 in *Gesammelte Abhandlungen*, **1**, 63–363, English edition, translated and edited F. Lemmermeyer and N. Schappacher, Springer Verlag, New York.

Hilbert, D. (1899) *Grundlagen der Geometrie* (*Festschrift zur Einweihung des Göttinger Gauss–Weber Denkmals*), Leipzig; revised 2nd ed. 1903, numerous subsequent editions.

Hilbert, D. (1900) Über den Zahlbegriff, *Jahresbericht der Deutschen Mathematiker-Vereinigung*, **8**, 180–184, in *Grundlagen der Geometrie*, 7th ed., Anhang VI, pp. 241–246.

Hilbert, D. (1901) Mathematische Probleme, *Archiv für Mathematik und Physik* 1, 44–63 and 213–237; reprinted in *Gesammelte Abhandlungen*, vol. 3, pp. 290–329; English trans. in Mathematical Problems: Lecture Delivered Before the International Congress of Mathematicians at Paris in 1900, trans. Mary F. Winston (1902), *Bulletin of the American Mathematical Society*, **8**, 437–479.

Hilbert, D. (1904) Über das Dirichletsche Prinzip, *Mathematische Annalen* 59, 161–186, in *Gesammelte Abhandlungen*, 3, 15–37.

Hilbert, D. (1906) Zur Varationsrechnung, *Mathematische Annalen*, **62**, 351–370, in *Gesammelte Abhandlungen*, 3, 38–55.

Hilbert, D. (1909) Hermann Minkowski, *Göttingen Nachrichten.* 72–101, rep. in *Mathematische Annalen*, **68** (1910), 445–471; rep. in *Gesammelte Abhandlungen* 3, 339–364.

Hilbert, D. (1917) Axiomatisches Denken, *Mathematische Annalen* **78**, 405–415, in *Gesammelte Abhandlungen* 3, 146–156.

Hilbert, D. (1922) Neubegründung der Mathematik. Erste Mitteilung, Abhandlungen Mathematische Seminar, Hamburg 1, 157–177, in *Gesammelte Abhandlungen* 3, 157–177.

Hilbert, D. (1926) Über das Unendliche, *Mathematische Annalen*, 95, 161–190, in *Grundlagen der Geometrie*, 7th ed, Anhang VIII, 262–288.

Hilbert, D. (1935) *Gesammelte Abhandlungen*, 3 vols. Springer Verlag, Berlin, 1932, 1933, 1935.

Hilbert D. and Bernays, P. (1934) *Die Grundlagen der Mathematik*, Springer Verlag, Berlin.

Hilbert, D. (1993) *Theory of Algebraic Invariants*, translated by Reinhard C. Laubenbachei; edited and with an Introduction by Bernd Sturmfels, Cambridge University Press, Cambridge.

Hocking, J. G. and Young, G. S. (1961) *Topology*, Addison Wesley, Reading, Mass.

Ilyashenko, Yu. (1995) *Concerning the Hilbert sixteenth problem*, *Advances with Mathematical Sciences*, 23, American Mathematical Society Translations, (2) 165, 1–190.

Itenberg. I. and Viro, O. (1996) Patchworking algebraic curves disproves the Ragsdale conjecture, *Mathematical Intelligencer*, 18.4, 19–28.

Jackson, A. (1999) The IHES at forty, *Notices of the American Mathematical Society*, 46.3, 329–337.

Kagan, B. (1903) Über die Transformation der Polyeder, *Mathematische Annalen*, 57, 421–424.

Kantor, J.-M. (1996) Hilbert Problems and Their Sequels, *Mathematical Intelligencer*, 18.1, 21–30.

Kleene S. C. (1952) *Introduction to Meta-mathematics*, North-Holland, Amsterdam.

Kleene, S. C. (1988) The Work of Kurt Gödel, pp. 48–73 in *Gödel's Theorem in Focus*, ed. S. G. Shanker, Routledge, London.

Kleiman, S. (1976) Rigorous foundation of Schubert's enumerative calculus, in Browder (1976) vol. 2, pp. 445–482.

Klein, C. F. (1882) *Über Riemanns Theorie der algebraischen Functionen und ihrer Integrale*, Teubner, Leipzig, English translation by F. Hardcastle (1893), *On Riemann's Theory of Algebraic Functions and their Integrals*, Macmillan, Cambridge, Mass.

Klein, C. F. (1884) *Vorlesungen über das Ikosaeder und die Auflösung der Gleichungen vom fünften Grade*. Teubner, Leipzig; English translation by G. G. Morrice (1888), *Lectures on the Icosahedron*, Dover reprint, 1956.

Klein, C. F. (1888) *Lectures on the Icosahedron*, English translation by G. G. Morrice (1956), Dover reprint, New York.

Klein, C. F. (1893) *The Evanston Colloquium Lectures*, Macmillan, New York.

Klein, C. F. (1926–27) *Vorlesungen über die Entwicklung der Mathematik im 19. Jahrhundert*, 2 vols., ed. R. Courant and O. Neugebauer, Springer Verlag, New York, Berlin Chelsea reprint, New York (1948).

Kneser, A. (1900) *Lehrbuch der Variations-rechnung*, Braunschweig.

Koebe, P. (1907), Über die Uniformisierung beliebiger analytische Kurven, *Göttingen Nachrichten*, 191–210.

Kolmogorov, A. N. (1930) *Grundlagen und Grundbegriffe der Wahrscheinlichkeitsrechnung*, Springer-Verlag, Berlin.

Krantz, S. C. (1990) Mathematical Anecdotes, *Mathematical Intelligencer*, **12**.4, 32–38.

Kreisel, G. (1976) Review of *Wittgenstein's lectures on the foundations of mathematics*, Cambridge 1939, *Bulletin of the American Mathematical Society*, **84**.1, 79–90.

Lacki, J. (2000) The early axiomatisations of quantum mechanics: Jordan, von Neumann and the continuation of Hilbert's programme, *Archive for History of Exact Sciences*, **54**(4), 279–318.

Lax, P. D. (1986) Mathematics and its Applications, *Mathematical Intelligencer*, **8**.4, 14–17.

Le Lionnais, F. (1971) *Great Currents of Mathematical Thought*, 1971 English translation of *Les Grands Courants de la Pensée mathématiques*, 2 vols (1947, 2nd ed. 1962) tr. R.A. Hall, H.G. Bergmann, C. Pinter, H. Kline.

Lorch, E. R. (1989) Mathematics at Columbia during Adolescence, *A Century of Mathematics in America* 3, ed. P. Duren, pp. 149–161, American Mathematical Society, History of Mathematics 3, Providence, Rhode Island.

Lützen, J. (1990) *Joseph Liouville, 1809–1882, Master of Pure and Applied Mathematics*, Springer-Verlag, New York.

Lützen J. (1999) Geometrising configurations. Heinrich Hertz and his mathematical precursors, in *The Symbolic Universe; Geometry and Physics 1890–1930*, pp. 25–46, ed. J. J. Gray, Oxford University Press, Oxford.

Machover, M. (1996) *Set theory, Logic, and Their Limitations*, Cambridge University Press, Cambridge.

Majer, U. (1997) Husserl and Hilbert on Completeness, *Synthese*, **110**.1, 37–56.

Matijasevich, Y. (1970) Enumerable sets are Diophantine, *Soviet Mathematics Doklady*, **11**, 354–357.

Matijasevich, Y. (1992) My collaboration with Julia Robinson, *Mathematical Intelligencer*, **14**.4, 38–45.

Maz'ya, V. and **Shaposhnikova, T.** (1998) *Jacques Hadamard, A Universal Mathematician*, American and London Mathematical Societies, History of Mathematics, **14**, Providence, Rhode Island.

Mehrtens, H. (1990) *Moderne—Sprache—Mathematik: eine Geschichte des Streits um die Grundlagen der Disziplin und des Subjekts formaler Systeme*, Suhrkamp Verlag, Frankfurt am Main.

Möhring, W. (1998) Hilbert's 18th problem and the Göttingen Town Library, *Mathematical Intelligencer* **20**.3, 43–44.

Montgomery, D. and **Zippin, L.** (1952) Small subgroups of finite-dimensional groups, *Annals of Mathematics*, **56**, 213–241.

Moore R. L. (1916) On the foundations of plane analysis situs, *Transactions of the American Mathematical Society*, **17**, 131–164.

Moore R. L. (1919) On the Lie–Riemann–Helmholtz–Hilbert Problem of the Foundations of Geometry, *American Journal of Mathematics*, **41**, 299–319.

Moore, G. H. (1982) *Zermelo's Axiom of Choice: Its Origins, Development and Influence*, New York.

Moore, G. H. (1988) The Emergence of First-Order Logic, pp. 95–15 in *Minnesota Studies in the Philosophy of Science*, ed. W. Aspray and P. Kitcher, **11**, University of Minnesota Press, Minneapolis.

Moulton, A. F. (1902) A simple non-Desarguesian Plane Geometry, *Trans. American Mathematical Society*, **3**, 192–195.

Nabonnand, P. (1999) The Poincaré–Mittag–Leffler Relationship, *Mathematical Intelligencer*, **21**.2, 58–64.

Nagata, M. (1962) On the fourteenth Problem of Hilbert, *Proceedings of the International Congress of Mathematicians, Edinburgh 1962*, 459–462.

Olesko, K. (1989) *Physics as a Calling; Discipline and Practice in the Königsberg Seminar for Physics*, Cornell History of Science Series, Cornell University Press.

Osgood, W. F. (1902) A Jordan curve of positive area, *Trans American Mathematical Society*, **4**, 107–112.

Padoa, A. (1900) Un nouveau système irréducible de postulats pour l'algèbre, *Compte Rendu du Deuxième Congrès international de Philosophie*, 249–256, Paris.

Padoa, A. (1902) Un nouveau système de définitions pour la géométrie euclidienne, *Compte Rendu du Deuxième Congrès international des mathématiciens'*, Paris.

Padoa, A. (1903) Le Problème no. 2 de M. David Hilbert, *L'Enseignement Mathématique*, **5**, 85–91.

Pais, A. (1982) *'Subtle is the Lord . . .' The Science and the Life of Albert Einstein*, Oxford University Press, Oxford.

Pais, A. (1986) *Inward Bound*, Clarendon Press, Oxford.

Parshall, K. H. and **Rowe, D. E.** (1994) *The Emergence of the American Mathematical Research Community; J. J. Sylvester, Felix Klein, and E. H. Moore*, American and London Mathematical Societies, **8**, Providence, Rhode Island.

Pasch, M. (1882) *Vorlesungen über neuere Geometrie*, Teubner, Leipzig.

Patterson, S. J. (1988) *An introduction to the theory of the Rieman zeta-function*, Cambridge University Press, New York.

Pfister, A. (1967) Zur Darstellurg definiter Funktionen als Summe von Quadraten, *Inventions Mathematicae*, **4**, 229–237.

Pfister, A. (1995) *Quadratic Forms with Applications to Algebraic Geometry and Topology*, London Mathematical Society Lecture Notes series 217, Cambridge University Press, Cambridge.

Picard, E. (1895) Sur une classe étendue des équations linéaires aux dérivées partielles dont tous les intégrales sont analytiques, *Comptes rendus* **121**, 12–14.

Poincaré, H. (1890) Sur les équations aux dérivées partielles de la physique mathématique, *American Journal of Mathematics*, **12**, 211–294, in *Oeuvres*, **9**, 28–113.

Poincaré, H. (1898) Sur les rapports de l'analyse pure et de la physique mathématique, *Verhandlungen des ersten Internationaler Mathematiker-Kongresses, Zürich, 1897*, F. Rudio, ed., Teubner, Leipzig, pp. 81–90.

Poincaré, H. (1901) *Électricité et Optique* (1st ed. 1890, 2nd ed., 1901) Gauthier-Villars, Paris, reprint Gabay, Paris, 1990.

Poincaré, H. (1906) Sur la dynamique de l'électron, *Rendiconti del circolo Matematico di Palermo*, **21**, 129–176, in *Oeuvres*, **9**, 494–550.

Poincaré, H. (1907) Sur l'uniformisation des fonctions analytiques, *Acta Mathematica*, **31**, 1–63, in *Oeuvres*, **4**, 70–139.

Poincaré, H. (1908) L'Avenir des mathématiques, *Rendiconti del Circolo Matematico di Palermo*, **26**, 152–168 (address to the International Congress of Mathematicians, Rome), reprinted in *Science et Méthode*, Flammarion, Paris (reprint 1916) 19–42, English translation in *Science and Method*, Dover Books, New York.

Pontrjagin, L. (1934) Sur les groupes topologiques compacts et le cinquième problème de M. Hilbert, *Comptes rendus de l'Académie des sciencs Paris,* **198**, 328–330.

Princeton Conference (1946) rep. in *A Century of Mathematics in America* **2**, pp. 309–334, ed. P. Duren, American Mathematical Society, History of Mathematics **2**, Providence, Rhode Island.

Radó, T. (1930) On Plateau's problem, *Annals. of Mathematics,* (2) **31**, 457–469.

Radó, T. (1926) Das Hilbertsche Theorem über den analytischen Charakter der Lösungen der partiellen Differentialgleichungen zweiter Ordnung, *Mathematische Zeitschrift,* **25**, 514–589.

Radó, T. (1930a) On Plateau's problem. *Annals of Mathematics,* (2) **31**, 457–469.

Radó, T. (1930b) The problem of the least area and the problem of Plateau. *Mathematische Zeitschrift,* **32**, 763–796.

Ragsdale, V. (1906) On the arrangement of the real branches of plane algebraic curves, *American Journal of Mathematics,* **28**, 377–404.

Regis, E. (1987) *Who got Einstein's Office?,* Addison-Wesley, Reading, Mass.

Reid, C. (1970) *Hilbert,* Springer Verlag, New York.

Reid, C. (1976) *Courant in Göttingen and New York,* Springer Verlag, New York.

Reingold, N. (1988) Refugee Mathematicians in the United States of America, 1933–1941: Reception and Reaction, *A Century of Mathematics in America* **1**, pp. 175–200, ed. P. Duren, American Mathematical Society, History of Mathematics **1**, Providence, Rhode Island.

Riemann, B. (1990) *Gesammelte Mathematische Werke, Wissenschaftliche Nachlass und Nachträge Collected Papers,* ed. R. Narasimhan, Springer-Verlag, New York.

Robb, A. A. (1911) *Optical Geometry of Motion: a New View of Theory of Relativity,* Cambridge University Press, Cambridge.

Robb, A. A. (1914) *A Theory of Time and Space,* Heffer, Cambridge.

Rodriguez-Consuegra, F. (1991) *The Mathematical Philosophy of Bertrand Russell: Origins and Development,* Birkhäuser Verlag, Boston and Basel.

Rowe, D. E. (1989) Klein, Hilbert, and the Göttingen Tradition, in K. M. Olesko, ed. *Osiris,* **5**, 186–213.

Rowe, D. E. (1999) The Göttingen response to general relativity and Emmy Noether's theorems, in *The Symbolic Universe; Geometry and*

Physics 1890–1930, pp. 189–233, ed. J. J. Gray, Oxford University Press, Oxford.

Rüdenberg, L. and **H. Zassenhaus** (eds.) (1973) *Hermann Minkowski–Briefe an David Hilbert*, Berlin/New York, Springer.

Sagan, H. (1994) *Space-filling curves*, Springer-Verlag, New York, Heidelberg.

Saul, M. (1999) Kerosinka: an episode in the history of Soviet mathematics, *Notices of the American Mathematical Society*, **46**, 1217–1220.

Scanlon, M. (1991) Who were the American postulate theorists?, *The Journal of Symbolic Logic*, **56**, 981–1002.

Schappacher, N. (1998) On the History of Hilbert's Twelfth Problem, in *Matériaux pour l'histoire des Mathématiques au XXe Siecle*, pp. 243–274, Société Mathématique de France.

Schneider, Th. (1934) Tranzendenzuntersuchungen periodischen Funktionen, I, *Journal für die reine und angewandte Mathematik*, **172**, 65–69.

Scholz, E. (1989) *Symmetrie—Gruppe—Dualität*, Birkhäuser, Boston and Basel.

Scholz, E. (1995) Hermann Weyl's purely 'infinitesimal geometry', *Proceedings of the International Congress of Mathematicians, Zürich 1994*, 1592–1603, Birkhäuser, Boston and Basel.

Scott, C. A. (1900) Report on the International Congress of Mathematicians in Paris, *Bulletin of the American Mathematical Society*, **7**, 57–79.

Sieg, W. (1998) Hilbert's programs: 1917–1922, *Bulletin of Symbolic Logic*, **5**, 1–44.

Sinaceur, H. (1991) *Corps et Modèles*, Vrin, Paris.

Sloane, N. J. A. (1998) The sphere packing problem, *International Congress of Mathematicians 1998*, **3**, 387–396.

Tarski, A. (1959) What is elementary geometry? pp. 16–29 in *The Axiomatic Method* (eds. L. Henkin, P. Suppes and A. Tarski) Studies in Logic and the Foundations of Mathematics, North-Holland, Amsterdam.

Toepell M.-M. (1985) On the origins of David Hilbert's *'Grundlagen der Geometrie'*, *Archive for History of Exact Sciences*, **36**, 329–344.

Toepell, M.-M. (1986) *Über die Entstehung von David Hilberts 'Grundlagen der Geometrie'*, Studien zur Wissenschafts-, Sozial-, und Bildungsgeschichte der Mathematik, **2**, Göttingen.

van Dalen, D. (1990) The war of the frogs and the mice, or the crisis of the *Mathematische Annalen*, *Mathematical Intelligencer*, **12.4**, 17–31.

van Dalen, D. (1999) *Mystic, Geometer, and Intuitionist, The Life of L. E. J. Brouwer* 1, *The Dawning Revolution*, Clarendon Press, Oxford.

Veblen, O. (1905) *Princeton lectures 'On the foundations of geometry'* , Chicago University Press.

Waltershausen, S. von (1856) *Gauss zum Gedächtniss*, Sändig, Leipzig.

Weber, H. (1893) *Lehrbuch der Algebra*, 3 vols', Teubner, Leipzig.

Weil, A. (1971) 'The future of mathematics' in Le Lionnais (1971), pp. 321–336.

Weil, A. (1984) *Number theory. An approach through history, from Hammurapi to Legendre*, Birkhäuser, Boston and Basel.

Westfall R. S. (1983) *Never at Rest; a Biography of Isaac Newton*, Cambridge University Press, Cambridge.

Weyl, H. (1944) David Hilbert and his Mathematical Work, *Bulletin of the American Mathematical Society*, **50**, 612–654, in *Gesammelte Abhandlungen* (1968), **4**, 130–172.

Wiles, A. (1995) Modular elliptic curves and Fermat's Last Theorem, *Annals of Mathematics*, **142**, 443–551.

Wright, J. E. (1906) The ovals of plane sextic curves, *American Journal of Mathematics*, 305–308.

Zygmunt, J. (1991) Mojzesz Presburgen: life and work, *History and Philosophy of Logic*, **12**, 211–223.

Index

This index contains no specific references to Hilbert himself, nor does it cover the text of his lecture (pp. 240–82).